Author:
Cindy Christianson
Science/Math Specialist

Illustrator:
Keith Vasconcelles

Editors:
Evan D. Forbes, M.S. Ed.
Walter Kelly, M.A.
Carol Amato, M.A.

Senior Editor:
Sharon Coan, M.S. Ed.

Art Direction:
Elayne Roberts

Product Manager:
Phil Garcia

Imaging:
Alfred Lau

Research:
Bobbie Johnson

Photo Cover Credit:
Images © PhotoDisc, Inc., 1994

Publishers:
Rachelle Cracchiolo, M.S. Ed.
Mary Dupuy Smith, M.S. Ed.

W9-AAJ-971

Hands-On Minds-On Science

Magnets
Early Childhood

Teacher Created Materials, Inc.
P.O. Box 1040
Huntington Beach, CA 92647
©1994 Teacher Created Materials, Inc.
Made in U.S.A.

ISBN-1-55734-612-7

Table of Contents

Table of Contents *(cont.)*

Introduction

What Is Science?

What is science to young children? Is it something that they know is a part of their world? Is it a textbook in the classroom? Is it a tadpole changing into a frog? Is it a sprouting seed, a rainy day, a boiling pot, a turning wheel, a pretty rock, or a moonlit sky? Is science fun and filled with wonder and meaning? What is science to children?

Science offers you and your eager children opportunities to explore the world around you and to make connections between the things you experience. The world becomes your classroom, and you, the teacher, a guide.

Science can, and should, fill children with wonder. It should cause them to be filled with questions and the desire to discover the answers to their questions. And, once they have discovered answers, they should be actively seeking new questions to answer.

The books in this series give you and the children in your classroom the opportunity to learn from the whole of your experience—the sights, sounds, smells, tastes, and touches, as well as what you read, write about, and do. This whole-science approach allows you to experience and understand your world as you explore science concepts and skills together.

What Are Magnets?

Magnets occur naturally in the ground. Some rocks in the ground are made of an iron ore called *magnetite*. They were called "magic stones" many centuries ago because it was discovered that a piece of this stone would point north and south while suspended from a string. A natural magnet is called a *lodestone,* which means "leading stone." When brought near iron items, it grabs them. Lodestones have north and south poles just as manufactured magnets do.

Using a hands-on, minds-on science approach to learning, your children will experience the wonderful world of magnets through the activities in this book. The activities have been divided into five main question sections: What is magnetism? How strong are magnets? What is a magnetic field? Will a magnet attract or repel? What are some uses of magnets? Each main question section offers children the opportunity to explore via discovery and simulation experiences the answers to these, as well as other, posed questions. Each question section also includes a child "do-along" book to carry a child's learning beyond classroom hours. Our desire is for you and your children to fall in love with learning while investigating the amazing world of magnets.

Scientific Method for Young Learners

The "scientific method" is one of several creative and systematic processes for proving or disproving a given question, following an observation. When the "scientific method" is used in the classroom, a basic set of guiding principles and procedures is followed in order to answer a question. However, real world science is often not as rigid as the "scientific method" would have us believe.

This systematic method of problem solving will be described in the paragraphs that follow.

Make an OBSERVATION.

The teacher presents a situation, gives a demonstration, or reads background material that interests children and prompts them to ask questions. Or children can make observations and generate questions on their own as they study a topic.

Example: Show children as many different pictures of magnets as you can find.

Select a QUESTION to investigate.

In order for children to select a question for a scientific investigation, they will have to consider the materials they have or can get, as well as the resources (books, magazines, people, etc.) actually available to them. You can help them make an inventory of their materials and resources, either individually or as a group.

Tell children that in order to successfully investigate the questions they have selected, they must be very clear about what they are asking. Discuss effective questions with your children. Depending upon their level, simplify the question or make it more specific.

Example: How many different shaped magnets are there?

Make a PREDICTION (Hypothesis).

Explain to children that a hypothesis is a good guess about what the answer to a question will probably be. But they do not want to make just any arbitrary guess. Encourage children to predict what they think will happen and why.

In order to formulate a hypothesis, children may have to gather more information through research.

Have children practice making hypotheses with questions you give them. Tell them to pretend they have already done their research. You want them to write each hypothesis so it follows these rules:

1. It is to the point.
2. It tells what will happen, based on what the question asks.
3. It follows the subject/verb relationship of the question.

Example: I think there are as many different shaped magnets as there are shapes.

4 | Develop a **PROCEDURE** to test the hypothesis.

The first thing children must do in developing a procedure (the test plan) is to determine the materials they will need.

They must state exactly what needs to be done in step-by-step order. If they do not place their directions in the right order, or if they leave out a step, it becomes difficult for someone else to follow their directions. A scientist never knows when other scientists will want to try the same experiment to see if they end up with the same results!

Example: By coloring a in variety of magnets children will become aware of the many different shapes magnets can be.

5 | Record the **RESULTS** of the investigation in written and picture form.

The results (data collected) of a scientific investigation are usually expressed two ways—in written form and in picture form. Both are summary statements. The written form reports the results with words. The picture form (often a chart or graph) reports the results so the information can be understood at a glance.

Example: The results of this investigation can be recorded on a data-capture sheet provided (page 24).

6 | State a **CONCLUSION** that tells what the results of the investigation mean.

The conclusion is a statement which tells the outcome of the investigation. It is drawn after the child has studied the results of the experiment, and it interprets the results in relation to the stated hypothesis. A conclusion statement may read something like either of the following: "The results show that the hypothesis is supported," or "The results show that the hypothesis is not supported." Then restate the hypothesis if it was supported or revise it if it was not supported.

Example: The hypothesis that stated "there are as many different shaped magnets as there are shapes" is supported (or not supported).

7 | Record **QUESTIONS, OBSERVATIONS**, and **SUGGESTIONS** for future investigations.

Children should be encouraged to reflect on the investigations that they complete. These reflections, like those of professional scientists, may produce questions that will lead to further investigations.

Example: How many different shapes are there?

Science-Process Skills for Young Learners

Observing

All information is acquired through the process of observing. That makes this skill the most fundamental of all the process skills. Children have been making observations all their lives, but they need to be made aware of how they can use their senses (seeing, hearing, smelling, tasting, touching) and their prior knowledge (life experiences) to gain as much information as possible from each experience. Teachers can develop this skill in children by asking them questions such as these:

Can you tell me what you see? What did you hear?

Can you point to what you are talking about?

Communicating

Humans have developed the ability to use language and symbols which allow them to communicate not only in the "here and now" but over time and space as well. The accumulation of knowledge in science, as in other fields, is due to this process skill. Of two main forms of communication, written and oral, young children have few problems with the oral. The written, however, often poses problems for teachers. Please keep in mind that "a picture speaks a thousand words." In an active-learning science environment, this process skill will include many visual displays (drawings and models), inventive spelling, much talking, sharing, and questioning. The style of questioning for the teacher should encourage children toward expressing their thoughts:

Can you draw a picture of what you just saw?

May I join your team and talk about what we just saw happen?

Can you write a sentence to go with the picture you just drew?

Classifying (*Comparing, Ordering, Categorizing*)

Once children are actively engaged in observing and communicating, they naturally begin to make comparisons, try to find order, and categorize what they are observing or doing. Comparing means noticing the similarities and the differences. Ordering includes finding patterns (smooth to rough, bright to dim, few to many) and sequencing events (time line or cycle). Categorizing includes trying to find connections between objects based on logical rationale (color, size, weight, etc.). Questions for encouraging this process skill would include the following:

How are these items alike? How are these objects different?

What happened first/second/last? What will happen next?

Can you find a way to show these objects are alike, but also different?

Can you show me three different groups you can make by using the same objects for all of your groups?

Organizing Your Unit

Designing a Science Lesson

In addition to the lessons presented in this unit, you will want to add lessons of your own, lessons that reflect the unique environment in which you live, as well as the interests of your children. When designing new lessons or revising old ones, try to include the following elements in your planning:

Question

Pose a question to your children that will guide them in the direction of the experiment you wish to perform. Encourage all answers, but you want to lead the children towards the experiment you are going to be doing. Remember, there must be an observation before there can be a question. (Refer to The Scientific Method, pages 5-6.)

Setting the Stage

Prepare your children for the lesson. Brainstorm to find out what children already know. Have children review books to discover what is already known about the subject. Invite them to share what they have learned.

Materials Needed for Each Group or Individual

List the materials each group or individual will need for the investigation. Include a data-capture sheet when appropriate.

Procedure

Make sure children know the steps to take to complete the activity. Whenever possible, ask them to determine the procedure. Make use of assigned roles in group work. Create (or have your children create) a data-capture sheet. Ask yourself, "How will my children record and report what they have discovered? Will they record numbers? Will they use different colored symbols to represent answers?" Let children record results orally, using a video or audio tape recorder. For written recording, encourage children to use the do-along books.

Extensions

Continue the success of the lesson. Consider which related skills or information you can tie into the lesson, like math, language arts skills, or something being learned in social studies. Make curriculum connections frequently and involve the children in making these connections. Extend the activity, whenever possible, to home investigations.

Closure

Encourage children to think about what they have learned and how the information connects to their own lives. Allow time for children to complete their do-along books.

Organizing Your Unit *(cont.)*

Structuring Child Groups for Scientific Investigations

Using cooperative learning strategies in conjunction with hands-on and discovery learning methods will benefit all the children taking part in the investigation.

Cooperative Learning Strategies

1. In cooperative learning, all group members need to work together to accomplish the task.
2. Cooperative learning groups should be heterogeneous.
3. Cooperative learning activities need to be designed so that each child contributes to the group and individual group members can be assessed on their performance.
4. Cooperative learning teams need to know the social as well as the academic objectives of a lesson.

Cooperative Learning Groups

Groups can be determined many ways for the scientific investigations in your class. Here is one way of forming groups that has proven to be successful in early childhood classrooms.

- **The Team Leader**—scientist in charge of following directions and setting up equipment.
- **The Physicist**—scientist in charge of carrying out directions (can be more than one child).
- **The Stenographer**—scientist in charge of recording all of the information.
- **The Transcriber**—scientist who communicates findings.

If the groups remain the same for more than one investigation, require each group to vary the people chosen for each job. All group members should get a chance to try each job at least once.

Using Centers for Scientific Investigations

Set up stations for each investigation. To accommodate several groups at a time, stations may be duplicated for the same investigation. Each station should contain directions for the activity, all necessary materials (or a list of materials for investigators to gather), a list of words (a word bank) which children may need for writing and speaking about the experience, and any data-capture sheets or needed materials for recording and reporting data and findings.

Model and demonstrate each of the activities for the whole group. Have directions at each station. During the modeling session, read the directions aloud while you carry out out the activity. When all children understand what they must do, let small groups conduct the investigations at the centers. You may wish to have a few groups working at the centers while others are occupied with other activities. In this case, you will want to set up a rotation schedule so all groups have a chance to work at the centers.

Assign each team to a station, and after they complete the task described, help them rotate in a clockwise order to the other stations. If some groups finish earlier than others, be prepared with another unit-related activity to keep children focused on main concepts. After all rotations have been made by all groups, come together as a class to discuss what was learned.

Do-Along Books

For each of the five magnet sections (What Is Magnetism? How Strong Are Magnets? What Is a Magnetic Field? Will It Attract or Repel? What Are Some Uses of Magnets?) there is a supplementary do-along book. Each book is designed so that children, in a hands-on/kinesthetic way, can create concrete images that illustrate concepts taught. Each book consists of six pages (three large pages, cut out and folded in half) containing simple sentences that correspond to the do-along book's page activity. The reason art work may not necessarily be "complete" on some pages is so children become more involved in the book's creation. On each page immediately following "Just the Facts," you will find specific directions for what is needed, as well as directions for completing that section's do-along book. For the general assembly of a do-along book, follow directions below:

1. Copy all large do-along pages; cut around outer rectangle's black line.

2. Fold pages in half on dashed line. Glue underside of "pages" together; allow to dry.

3. Layer book in order: front cover and page 1, pages 2 and 3, pages 4 and 5, etc.

4. Make certain all left side edges are flush. Staple book together near top, middle, and bottom of left side edge. (Optional: Use hole puncher/string method.)

5. Write "title" of book on cover, along with child's name (or have child write own name). Have child decorate front cover after overall concepts (pages) have been introduced.

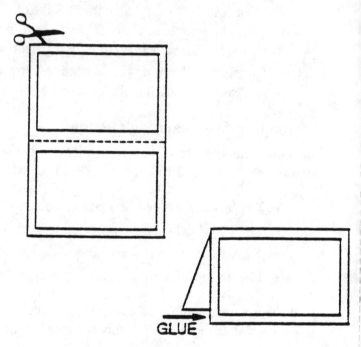

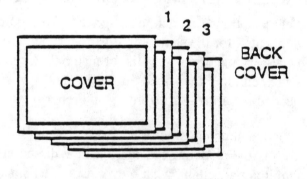

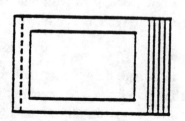

Just the Facts

Thousands of years ago, the ancient Greeks and Romans were aware of magnetism. They saw that a mysterious black stone, found in Asia Minor, had a force that made certain metals cling. They also noticed that when a piece of this stone was held by thread, one end always pointed to the north. Such a stone could be used as a compass for indicating direction. The Greeks and Romans had difficulty in explaining these strange happenings. They believed that the power of the black rock was caused by a supernatural force. From this, many legends arose in the ancient world. One legend told about the shepherds from the Greek Province of Magnesia. The shepherds carried wooden staffs and covered the ends with iron so they would not wear out so quickly on the rocky ground. As one shepherd tended to his flocks on Mount Ida, he noticed that some of the tiny stones in the soil clung to the iron tip of his staff. He had a hard time pulling the stones off. He called these stones "magnets" after his homeland. Other Greeks called this rock "magnet stone," or magnes, and thought it had magical properties.

The ancient Chinese also discovered these black stones and learned that when they hung an elongated stone of this type from a string, it always pointed north and south. They called these stones "The Stone That Picks Up Iron" and *Tchi-nan,* the chariot of the south. Legend says it guided many caravans across the Tatary grasslands of Asia.

In reality, these stones are not stones at all, but pieces of iron ore.

During the Middle Ages, people were still fascinated by magnets, but they had very little idea how they worked. Even those with scientific knowledge had strange ideas about the powers of magnets.

Finally, at the end of the 16th century, an English scientist named William Gilbert began to examine the way magnets work. He suggested that the earth itself was a huge magnet, with poles like an ordinary magnet.

The mineral was generally known in English as "lodestone." The name lodestone signifies "leading stone" and refers to one of the first uses of magnetism—the compass.

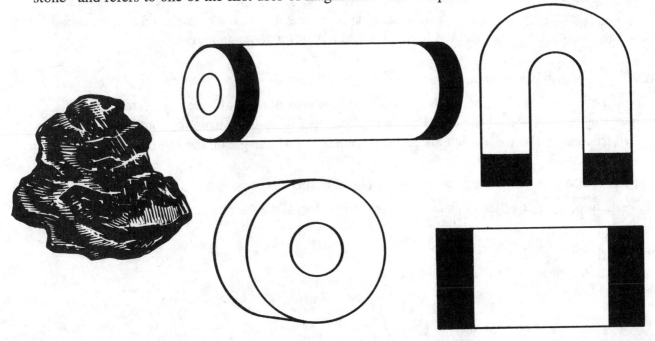

Let's Look at Magnets

(For general book-making directions, see page 10.)

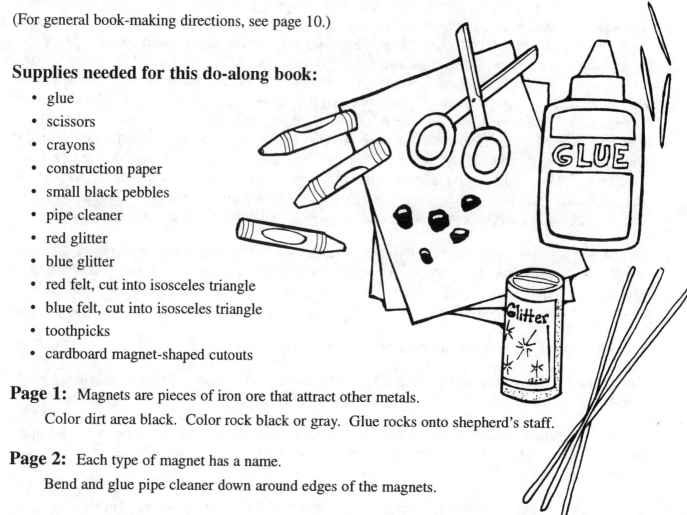

Supplies needed for this do-along book:

- glue
- scissors
- crayons
- construction paper
- small black pebbles
- pipe cleaner
- red glitter
- blue glitter
- red felt, cut into isosceles triangle
- blue felt, cut into isosceles triangle
- toothpicks
- cardboard magnet-shaped cutouts

Page 1: Magnets are pieces of iron ore that attract other metals.

Color dirt area black. Color rock black or gray. Glue rocks onto shepherd's staff.

Page 2: Each type of magnet has a name.

Bend and glue pipe cleaner down around edges of the magnets.

Page 3: A magnet has two poles.

Using fingertip, place thin layer of white glue over the surface area of each end of the magnets. Gently shake red glitter over the north pole and blue glitter over south pole.

Page 4: The earth is a magnet. It has two poles, too.

Color the water area of the earth blue. Color the land area of the earth green. Paste a toothpick pointing up from the north pole. Paste a toothpick pointing down from the south pole. Past a red felt triangle on the north pole toothpick. Paste a blue felt triangle on the south pole triangle.

Page 5: A compass uses a magnet. It always points north.

Paste a toothpick on the compass to act as the needle. Make the pointed end face north.

Page 6: Magnets have different shapes.

Match the shape to the magnet and glue on.

12

Let's Look at Magnets (cont.)

Each type of magnet has a name.

2

Magnets are pieces of iron ore that attract other metals.

1

Let's Look at Magnets (cont.)

4

The earth is a magnet. It has two poles, too.

A magnet has two poles.

3

Let's Look at Magnets (cont.)

9

Magnets have different shapes.

A compass uses a magnet. It always points north.

5

It Is Simply Attraction

Question

What is magnetism?

Setting the Stage

- Brainstorm with children about magnetism to find out what they already know.
- Set up a magnets display showing how each magnet is used.

Materials Needed for Each Group

- bar magnet
- paper
- paper clip
- string

Procedure

1. Have children wrap the magnet in paper so that it is hidden.
2. Then ask children, or help them, to attach the string to the paper clip and lay it on a flat surface.
3. Tell children to pass the paper (with the hidden magnet inside) over the paper clip and watch the paper clip move and drag the string with it.
4. Ask children how and why this happens.

Extension

Have children try passing the hidden magnet over the paper clip to make the string wiggle.

Closure

Have children complete page 1 of the *Let's Look at Magnets* do-along book.

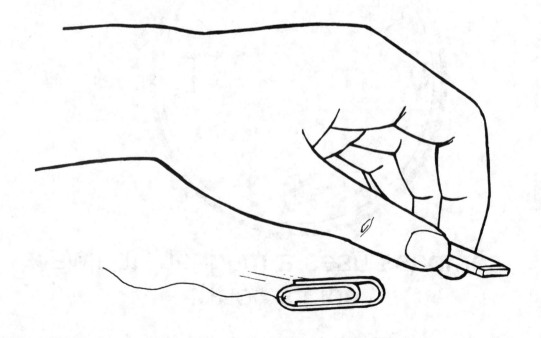

Name That Magnet

Question
> How many types of magnets are there?

Setting the Stage
- Brainstorm with children about the shapes of different magnets.
- Set up a display of the various types of magnets.

Materials Needed for Each Group
- horseshoe magnet
- bar magnet
- rod magnet
- U-shaped magnet
- block magnet
- assorted magnet pictures
- colored markers or crayons
- scissors
- stapler
- mini-book of magnets (page 18), one per child
- data-capture sheet (page 19), one per child

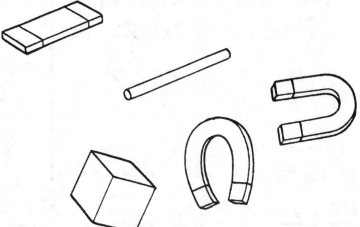

Procedure
1. Display a large picture of all the magnets for your class and identify them with your children.
2. Use teacher information as the basis when explaining what a magnet is.
3. Have children examine each of the magnets that will be used in this experience. You can do this by letting your children hold and play with the magnets.
4. Allow children to feel the "strength" of each magnet by placing several magnets close together.
5. Have children complete their data-capture sheets. This will help them recognize different types of magnets.

Extension
> Have children bring assorted magnets from home to share with the class.

Closure
> Have children complete page 2 of the *Let's Look at Magnets* do-along book.

Name That Magnet (cont.)

Mini-Book of Magnets

Color the different magnets. Cut along the dark lines and staple together to make a mini-book of magnets.

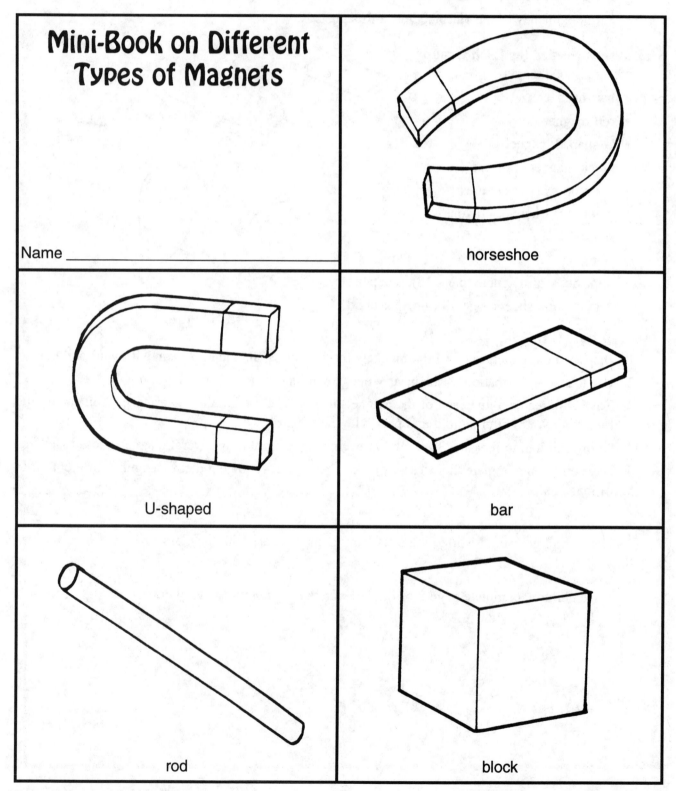

Mini-Book on Different Types of Magnets

Name _____

horseshoe

U-shaped

bar

rod

block

Name That Magnet *(cont.)*

Find and color the different types of magnets.

Magnetic Poles

Question
What is a magnetic pole?

Setting the Stage
Tell children some parts of a magnet are stronger than other parts. Tell children the strongest part of a magnet is called a pole. Every magnet has two poles, often called the north pole and the south pole.

Materials Needed for Each Individual
- bar magnet
- horseshoe magnet
- crayons
- data-capture sheet (page 21)

Procedure
1. Give each child a copy of the data-capture sheet. Then give your children the following instructions.
2. A bar magnet has two poles, one at each end. Can you find the bar magnet? Color one pole blue and one pole red.
3. A horseshoe magnet has two poles, too, one at each end. Can you find the horseshoe magnet? Color one pole blue and one pole red.
4. Have children complete their data-capture sheets after you have explained what they need to do.

Extension
Ask your children the following questions as an oral quiz and to stimulate discussion:

1. The strongest parts of a magnet are the _____.
2. Every magnet has _____ poles.
3. The poles are called the _____ pole and the _____ pole.

Answer Key
1. poles
2. two
3. north, south

Closure
Have children complete pages 3 and 4 of the *Let's Look at Magnets* do-along book.

Magnetic Poles

Color the magnetic poles with blue and red.

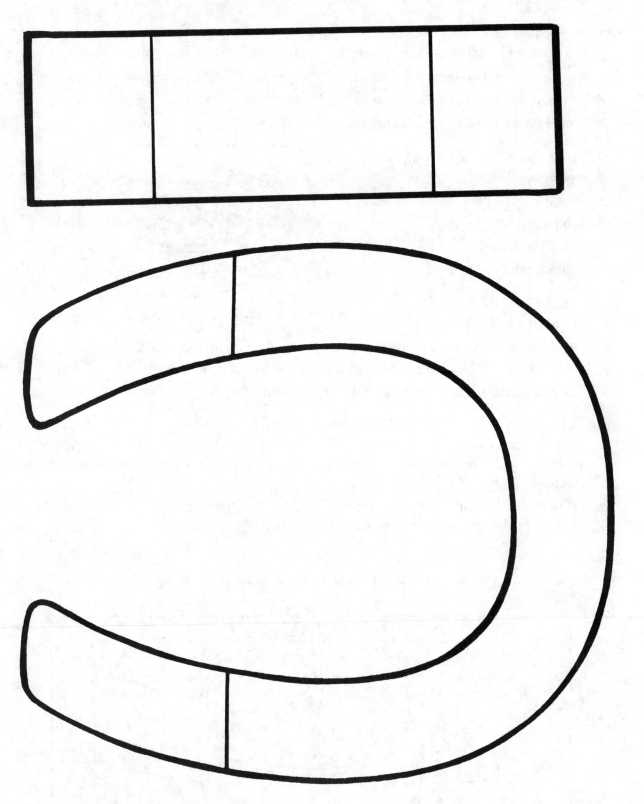

Compasses

Question

How does a compass work?

Setting the Stage

- Ask children if they know how a compass works.
- Ask children if they know the main directions of north, south, east, and west.
- Ask children if they can tell directions by the sun in the morning and evening.
- Locate the four main directions in the room for them.

Materials Needed for Each Group

- sewing needle
- flat cork or flat piece of styrofoam
- dish pan
- horseshoe magnet
- pictures of real compasses

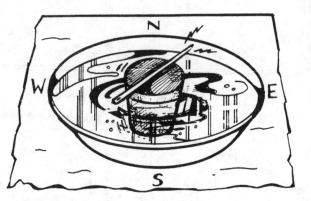

Procedure (*Child Instructions*)

1. Fill the pan with some water.
2. Stroke the needle on the magnet in one direction, about 50 times, to magnetize it.
3. Lay the needle on top of the cork or styrofoam and watch what happens.
4. Slowly turn the needle around and watch the movement again.

Extension

- Ask children to observe what happens. The needle should point to the north, just like a real compass does. The needle is a simple compass. In its natural position, the needle is pointing toward the magnetic north and south.
- Explain that no matter which way you move it, it will always turn and point to the north. This is the principle behind the compass a navigator would use.

Closure

Have children complete page 5 of the *Let's Look at Magnets* do-along book.

Shapes of Magnets

Question

How many different shaped magnets are there?

Setting the Stage

Show children pictures of as many different kinds of magnets as you can find.

Materials Needed for Each Individual

- blue, brown, purple, and red crayons
- data-capture sheet (page 24)

Procedure

1. Give each child a copy of the data-capture sheet. Then tell your children the following: Magnets can be many shapes. Some magnets are round. Other magnets are shaped like a long bar, or rectangle.
2. Tell children to color the horseshoe magnet brown.
3. Tell children to color the round magnet purple.
4. Tell children to color the square magnet red.
5. Tell children to color the bar magnet blue.

Extension

Have children make a classroom display with their data-capture sheets.

Closure

Have children complete page 6 of the *Let's Look at Magnets* do-along book.

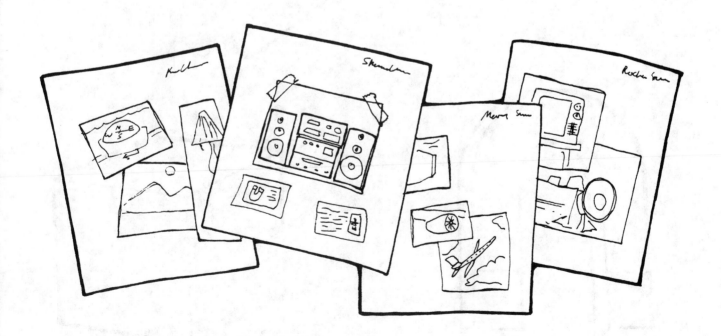

Shapes of Magnets *(cont.)*

Color the different magnets.

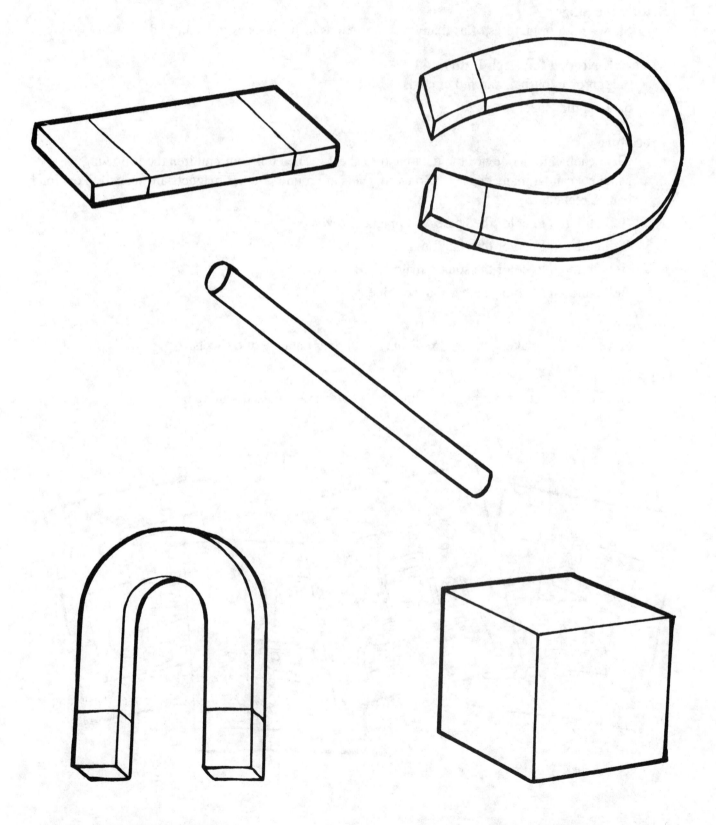

Just the Facts

All materials in the earth are made up of small particles called molecules. Iron is made of molecules and each little molecule of iron is a magnet. When the molecules of a piece of iron are mixed, the poles point in every possible direction and the iron is not magnetized. When the molecules get lined up so that the south poles point in one direction and the north poles in the other, the iron is magnetized. The magnet is strongest at its two poles, and weakest in the middle.

Magnets of soft iron that can be magnetized easily, but lose their magnetism easily, are called "temporary" magnets. Iron will keep its magnetism for quite a long time, but will lose it more quickly than a piece of hard steel. For this reason, most bar and horseshoe magnets are made from steel. When a magnet can keep its magnetism for a long time, it is often referred to as a "permanent" magnet.

In the late 1950s, scientists developed a new kind of steel called *alnico*. Alnico is made of iron, carbon, aluminum, nickel, and cobalt, and is far stronger than magnetized steel. Magnets made of alnico can lift up to 1,000 times their own weight.

The earth itself is a huge permanent magnet. It is made up of four main parts: an inner core, an outer core, a mantle, and the crust. Scientists think that the inner core is solid, and made of iron and nickel. The outer core is like a thick pudding. The movement of the inner core inside the outer core makes a magnetic field. The earth's magnetic field acts between its magnetic north pole and magnetic south pole.

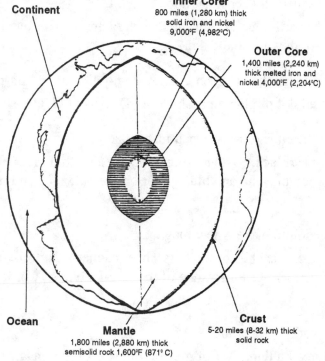

Continent

Inner Corer
800 miles (1,280 km) thick
solid iron and nickel
9,000°F (4,982°C)

Outer Core
1,400 miles (2,240 km)
thick melted iron and
nickel 4,000°F (2,204°C)

Ocean

Mantle
1,800 miles (2,880 km) thick
semisolid rock 1,600°F (871° C)

Crust
5-20 miles (8-32 km) thick
solid rock

Scientists believe that all the planets and stars have magnetism. In fact, they believe that the whole universe has this invisible force. Our satellites and space probes are finding out more information about magnetic fields in space.

Permanent magnets are part of everyday life. They are at work in electric motors, stereo speakers, telephones, and televisions.

Magnetic Strength

(For general book-making directions, see page 9.)

Supplies needed for this do-along book:
- crayons
- construction paper
- glue
- red glitter
- blue glitter
- paper clips
- tacks
- small nails
- scissors

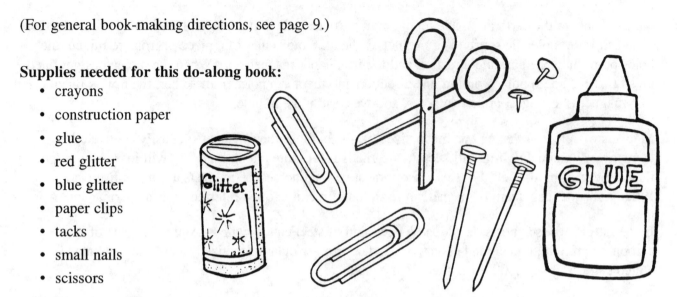

Page 1: Some magnets are very strong. They can lift cars.

Color the magnet gray. Cut out a piece of red construction paper in the shape of a car and glue under the magnet.

Page 2: The earth is a magnet. It keeps the moon in orbit.

Color the water area of the earth blue. Color the land area green. Cut out a circle from gray construction paper and glue down as the moon.

Page 3: Atoms line up in magnetized objects. They do not line up in non-magnetized ones.

Using fingertip, place small dot of glue on each atom in the nail. Gently shake red glitter on the atoms.

Page 4: Objects stick to the ends of rod-shaped magnets.

Color the rod-shaped magnet black. Using a fingertip, spread a thin layer of white glue over each pole. Gently sprinkle red glitter on one end and blue glitter on the other. Then glue paper clips, tacks, and small nails to each end.

Page 5: Rub in one direction to make a new magnet.

Spread a thin layer of glue on the ends of the horseshoe magnet. Carefully sprinkle red glitter on one end and blue glitter on the other. Color the rest of the magnet black. Color the nail yellow. Color the force lines blue and red.

Page 6: A new magnet works for a while.

Color the nail yellow. Spread a thin layer of glue on the ends of the magnet. Sprinkle blue glitter on one end and red glitter on the other. Glue paper clips and a safety pin to the end of the nail.

Magnetic Strength (cont.)

The earth is a magnet. It keeps the moon in orbit.

2

Some magnets are very strong. They can lift cars.

1

Magnetic Strength *(cont.)*

4

Objects stick to the ends of rod-shaped magnets.

Atoms line up in magnetized objects. They do not line up in non-magnetized ones.

3

Magnetic Strength (cont.)

9

A new magnet works for a while.

- -

Rub back and forth to make a new magnet.

5

Paper Clip Attraction

Question

How many paper clips can cling to a magnet?

Setting the Stage

Hold up a large horseshoe magnet and have children guess how many paper clips will be attracted to it.

Materials Needed for Each Group

- bar or horseshoe magnet
- box of paper clips
- plastic cup
- colored markers or crayons
- copy of Magnetic Strength (page 31), one per child
- copy of Magnetic Train (page 32), one per child

Procedure

Test 1

1. Have children empty the boxes of paper clips into their plastic cups.
2. Ask each child to dip the magnet into the cup to see how many paper clips it picks up. Give them three tries to get as many as they can. Then, have them record their results on page 31.

Test 2

1. Have children see how many paper clips they can hang from the end of one magnet.
2. Have children pick up one paper clip with the magnet.
3. Then ask them to take a second paper clip, touch it to the end of the first one, and try to get it to stick.
4. Have children continue doing this until no more paper clips are attracted by the magnet. Have them record their results on page 32.

Extensions

- Have children repeat this experience trying other magnetic materials to see how many things can be attracted at one time.
- Tell children the paper clips are only magnetized with the big magnet. If you take the paper clips off, they lose their magnetism. Explain to them the paper clips, then, are an example of a temporary magnet.

Closure

Have children complete pages 1 and 2 of the *Magnetic Strength* do-along book.

Paper Clip Attraction *(cont.)*

Magnetic Strength

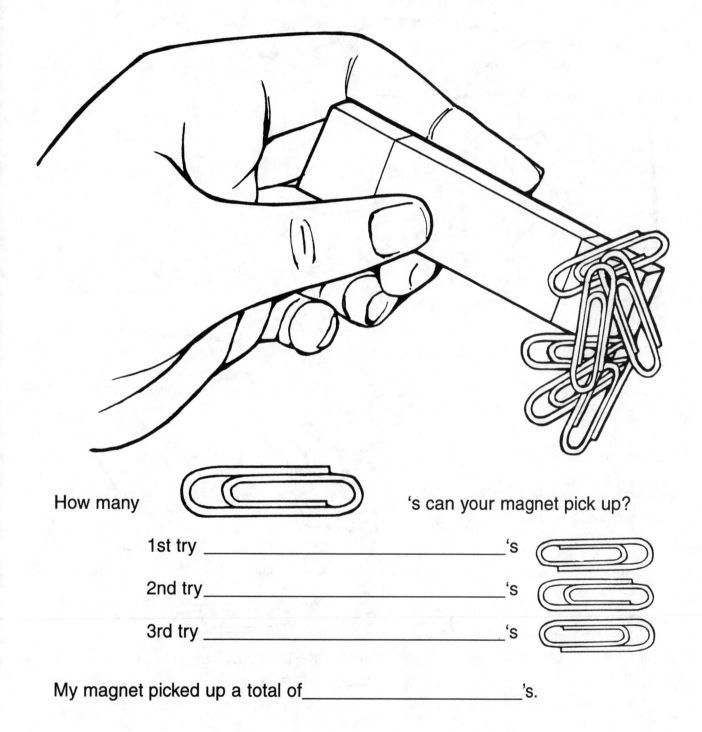

How many _____ 's can your magnet pick up?

1st try _____'s

2nd try _____'s

3rd try _____'s

My magnet picked up a total of_____'s.

Paper Clip Attraction (cont.)

Magnetic Train

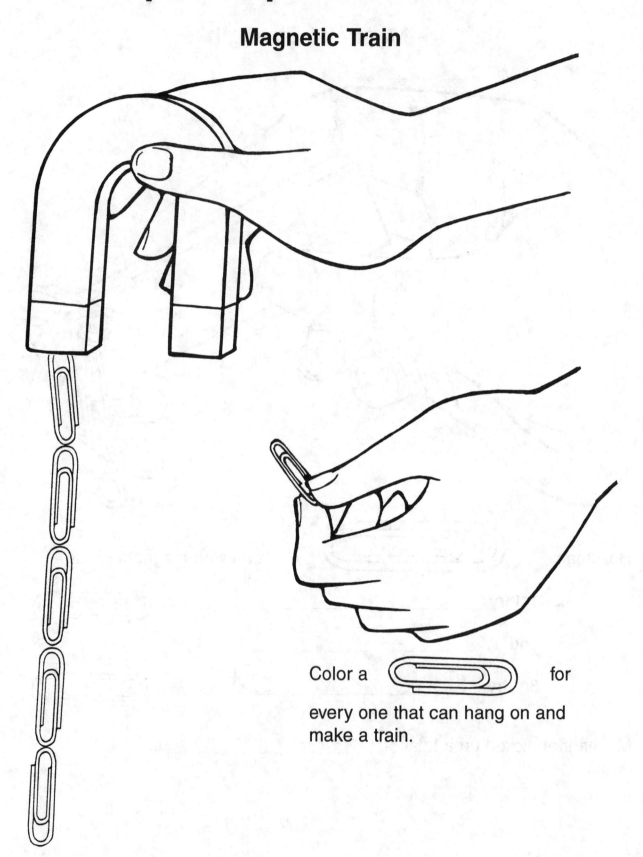

Color a [paper clip] for every one that can hang on and make a train.

Line Up

Question
Can you strengthen and weaken the magnetism in an object?

Setting the Stage
Ask children if they think they can pick up a smaller nail (or iron washer) with a larger nail.

Materials Needed for Each Group
- nail about 1 or 2" (2.5 or 5 cm) long
- small nails
- washers
- bar magnet

Procedure
1. Have children rub the nail against the pole of the magnet. The nail will now be magnetized.

2. Ask them to use the nail to pick up several small nails and washers, which it will do by magnetic attraction.

3. Ask children to bang the larger nail against a hard object a few times.

4. Then have them try to pick up the small nails again. Ask them to explain what happened. The magnetism of the nail should have become much weaker.

Extension
Explain to children materials differ in the ways their atoms can be lined up. Atoms do not line up in non-magnetic items, but in magnetic material, they line up very easily. Banging the magnet helps to shift the atoms so that they lose their lined-up condition.

Closure
Have children complete page 3 of the *Magnetic Strength* do-along book.

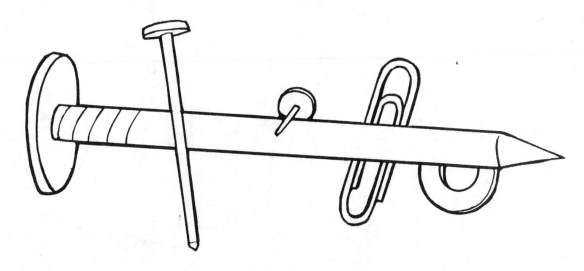

Tug of War

Question

Where do objects stick on a rod-shaped magnet?

Setting the Stage

Rattle a box of washers and ask children how to pick up most of them at one time.

Materials Needed for Each Group

- rod-shaped alnico magnet
- small nail
- small box of washers or brads

Procedure

1. Ask children to bring the head of a small nail just underneath the middle of the rod-shaped alnico magnet and let go. The nail tends to jump to the end of the magnet and stick there.

2. Then ask them to dip the same magnet into a pile of washers or brads. They will stick mainly to the ends of the magnet. A tiny bridge of iron objects will form from one end of the magnet to the other.

Extension

Explain to children this experience shows that the ends of the magnet have greater magnetic pull than the middle. The places of greatest magnetic strength in a magnet are the magnetic poles.

Closure

Have children complete page 4 of the *Magnetic Strength* do-along book.

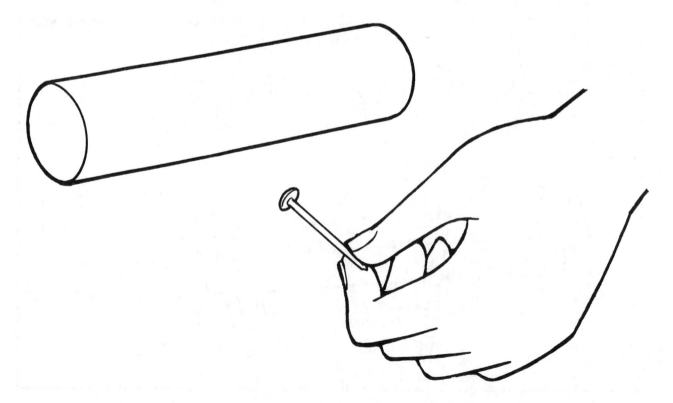

Magnet Magic

Question

Can you make a magnet from something that is not magnetic?

Setting the Stage

Ask children if it is possible to make a magnet from something that is not magnetic.

Materials Needed for Each Group

- bar magnet
- large nail
- iron tacks

Procedure

1. Ask children to try touching the tacks together to be certain that they are not magnetic.
2. Then ask them to stroke the nail on the magnet, moving in the same direction, about 50 times.
3. Now ask them to try picking up the tacks with the nail.

Extension

Explain to children the atoms inside the nail are randomly distributed with both positive and negative charges. The charges are then caused to align in one direction, causing the magnetism.

Closure

Have children complete page 5 of the *Magnetic Strength* do-along book.

Magnetic Power

Question
How long will a magnet hold its magnetic power?

Setting the Stage
Let children know that they will get a chance to make a magnet and to find out how strong it is.

Materials Needed for Each Group
- a strong magnet
- steel nail or wire
- small paper clips or safety pins

Procedure
1. Have children use the magnet and stroke the nail or wire, in the same direction, at least 60 times.
2. Ask them to use the magnet to pick up the paper clips or safety pins.
3. Then ask them to estimate how long their new "magnet" will hold.

Extension
Ask children to see if they can magnetize other types of objects in the same way.

Closure
Have children complete page 6 of the *Magnetic Strength* do-along book.

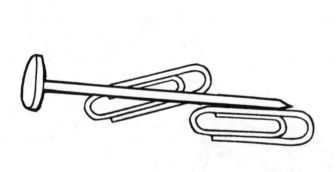

Just the Facts

Look at a magnet. Do you see anything around it? Do you feel anything when you move your hand through the air near the magnet? You see and feel nothing, but there is something around the magnet. This "something" is not like air. It cannot be trapped in a bottle, and it does not have weight. This "something" is, in fact, a magnetic field. It is the area where the force of a magnet acts or can be felt when it is held near another magnetized object.

The magnetic field around a magnet has no weight at all. It passes right through objects. If you hold magnetic materials near a very strong magnet, you can feel a force pulling the materials towards the magnet. This force is difficult to feel in weaker magnets.

A magnetic field is made up of what scientists called magnetic lines of force. These lines are also invisible. They are closed curves that leave the north-seeking pole of the magnet, travel through the air to the south-seeking pole, and return inside the magnet to the north-seeking pole again. The lines of force are bunched together and concentrated at the poles, and are spread farther apart and less concentrated at the middle of the magnet. That means they are strongest at the magnet's poles and weakest in the center. The lines of force never cross each other.

A magnet's field is always interacting with other magnetic fields. For the magnetic force to be observed, there must be an interaction between the magnet and other materials, such as soft iron. All substances display magnetic properties, but most of them only show these to a very small degree. Low levels of magnetism are usually detected only with highly sophisticated instruments.

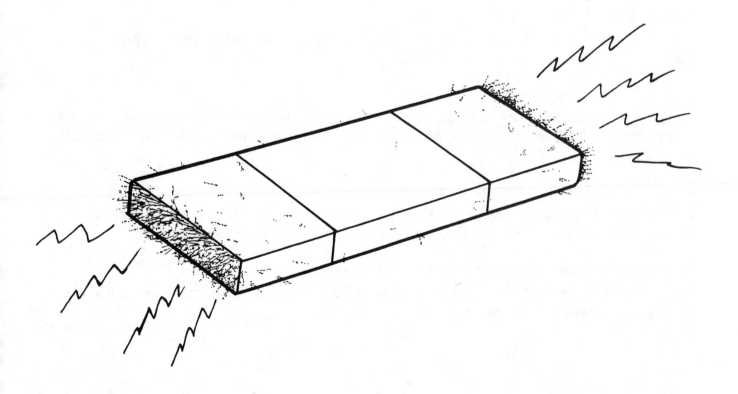

Magnetic Fields

(For general book-making directions, see page 10.)

Supplies needed for this do-along book:
- crayons
- glue
- scissors
- brown candy toppings
- black glitter
- silver glitter
- red glitter
- paper clips
- construction paper

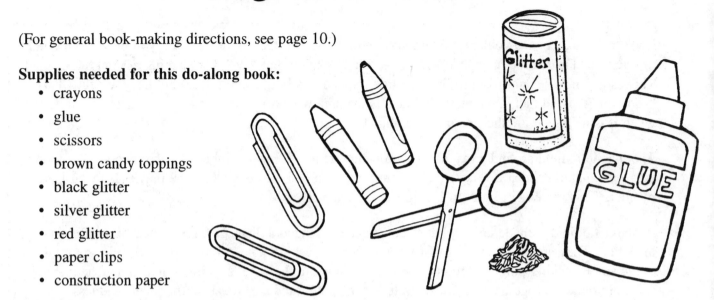

Page 1: The space around a magnet is called a magnetic field.

Color the bar magnet black. Color one pole blue and one pole red. Using a fingertip, spread a thin layer of glue around the magnet, leaving a 1/2" (1.25 cm) gap all around the magnet. Sprinkle silver glitter over the glue to represent iron filings.

Page 2: Magnetic forces work through the air.

Color bar magnets black. Using a fingertip, spread glue over the poles and in the area between the bars. Then lightly sprinkle some silver glitter over the area to represent iron-filings being attracted through the air.

Page 3: Magnetic forces work through sand.

Color the ground tan or gold to represent sand. Color the magnet black. Glue a paper clip onto the sand.

Page 4: Magnetic forces work through water.

Color the magnet black. Using a fingertip, lightly spread some glue on the N pole. Sprinkle red glitter over the glue. Cut a piece of blue construction paper in the shape of a small puddle. Glue a paper clip onto the puddle.

Page 5: Magnetic forces work through solid objects.

Color the rod magnet black. Glue on a piece of construction paper, cut like a square, so that one side overlaps the end of the rod magnet. Glue a paper clip so that it sticks out from the square.

Page 6: Magnets can make objects move without touching them.

Color the table brown. Color the buildings different colors. Paste on a construction paper cutout of a car. Paste on a cutout of a horseshoe magnet.

Magnetic Fields *(cont.)*

Magnetic forces work through the air.

2

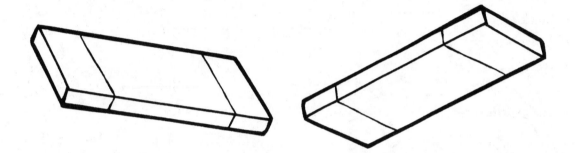

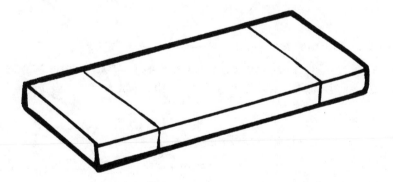

The space around a magnet is called a magnetic field.

1

Magnetic Fields *(cont.)*

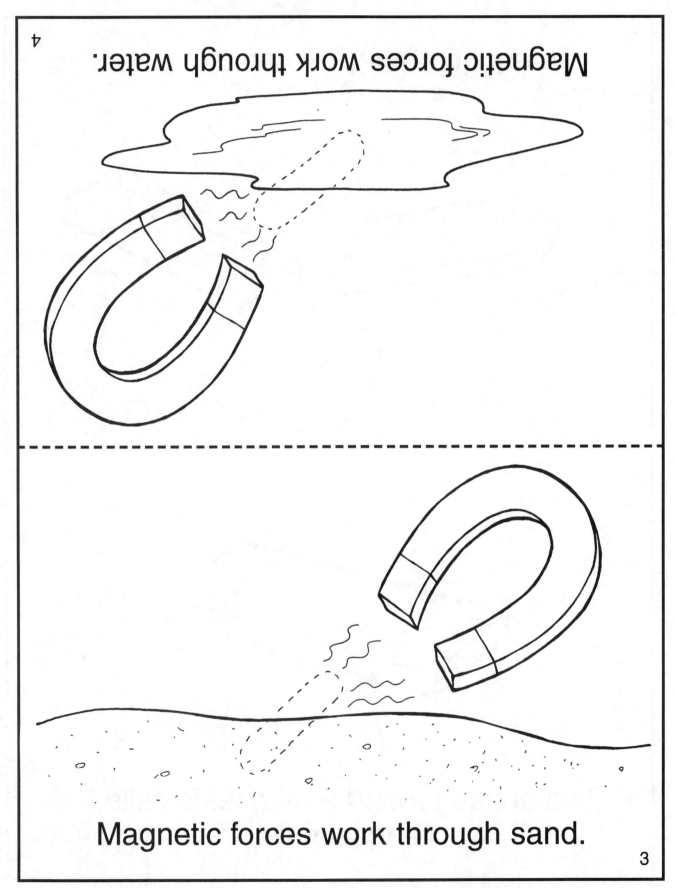

4.

Magnetic forces work through water.

Magnetic forces work through sand.

3

Magnetic Fields (cont.)

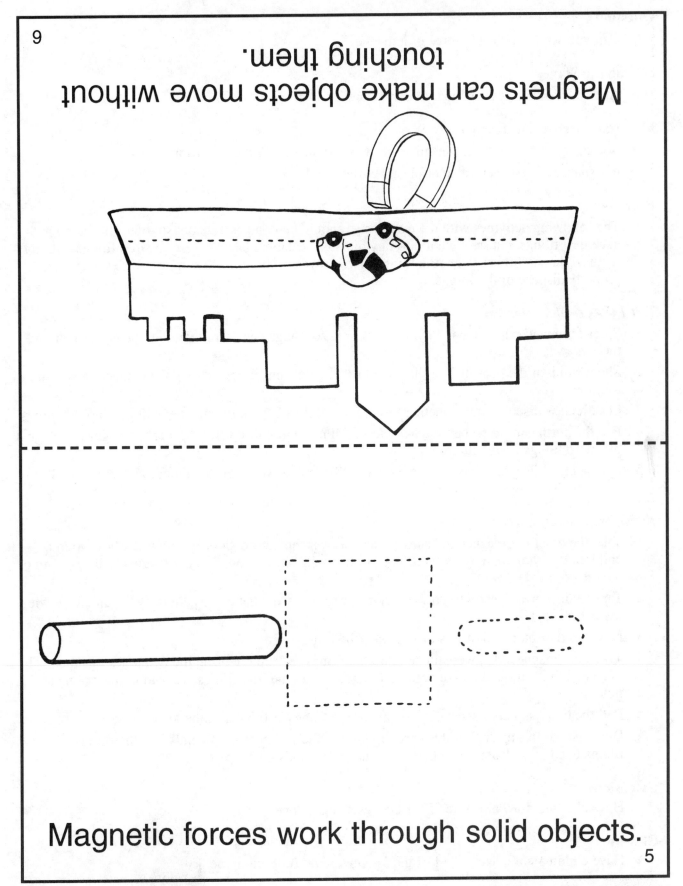

9

Magnets can make objects move without touching them.

Magnetic forces work through solid objects.

5

The Invisible Force

Question

Can you see the invisible force of a magnet?

Setting the Stage

Ask children if it is possible to see a magnet's force.

Materials Needed for Teacher

- iron filings
- newspaper
- two bar magnets of the same size (with their poles marked)
- petri dish or plastic coffee can lid

Procedure

The following activities with the iron filings should be done as a demonstration. If you do not have iron filings, cut steel wool into tiny bits in a sealable plastic bag so the iron filings will not scatter. Be careful not to get the filings in your eyes or mouth. Be sure to use newspaper to properly dispose of the filings.

Test 1 (Wear safety glasses)

1. Spread the newspaper down on your work area so filings can be rolled up in the newspaper and thrown away.
2. Show children the petri dish or plastic lid filled with iron filings. (Keep the magnet away from it for now.)
3. Lead a class discussion on whether or not your children think the magnet will pick up the filings.
4. Dip the north end of the bar magnet into the filings and pull up slowly. (The stronger the magnet, the more filings it will attract.)
5. Lead a class discussion about what is in the filings that would attract them to the magnet. Then place the bar magnet with the filings to the side.

Test 2 (Wear safety glasses)

1. Take the other bar magnet and slowly move it over the petri dish or plastic lid. Show your children that you are moving the filings around without touching the bar magnet to the bottom of the dish or lid.
2. Tap the dish or lid gently to erase previous patterns; see what other patterns you can create with the bar magnet.
3. Now dip the south end of this magnet into the filings.
4. Take both magnets that were dipped into the filings and slowly bring the two poles together. The magnetic fields around unlike poles join together and the iron filings are held in the magnetic field.
5. Pull the two magnets a short distance apart; the filings will hang in the air.
6. Demonstrate this again, but this time dip both of the magnets' north ends into the filings. The filings bend away from each other. The magnetic fields push apart.

Extension

Have children draw a picture of the two tests being done.

Closure

Have children complete page 1 of the *Magnetic Fields* do-along book.

42

Magnetism in the Air

Question
Does the magnetic force work through the air?

Setting the Stage
Ask children what their initials are and have them write them down on a piece of paper. Tell them they will be playing a game with the magnets to find their initials.

Note to the teacher: Help those who do not know what "initials" are or how to write them.

Materials Needed for Each Group
- construction paper
- rod or bar magnets
- small horseshoe or U-shaped magnets
- string
- pencils
- scissors
- paper clips
- large box

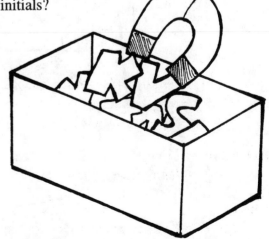

Procedure
1. Tie a small horseshoe or U-shaped magnet on the end of a string and then attach to a pole or stick to create a "fishing pole."
2. Outline each child's initials in large block letters on a piece of construction paper. Have children cut these letters out with scissors.
3. Then have children put a paper clip on each of their initials and place them in the large box.
4. Divide your class into several groups with about five or six children in each group.
5. Let each group "fish" for their initials in the box, using the magnetic fishing pole.
6. After each group has caught their initials, ask children if they can answer the main question: Does the magnetic force work through the air?

    If they cannot answer the question, ask them this: If magnetic force had not worked through the air, would you have caught your initials?

Extension
Brainstorm with your class how many other ways there are for magnets to work in the air.

Closure
Have children complete page 2 of the *Magnetic Fields* do-along book.

Buried Treasure

Question
Will magnets work through sand?

Setting the Stage
Let children know that they will have the chance to work in the sandbox or sandtable area.

Materials Needed for Each Group
- a sandbox or sandtable
- an assortment of magnetic objects (key, nail, paper clip, screw, nail, small cars and trucks, etc.)
- an assortment of magnets
- data-capture sheet (page 45), one per child

Procedure
1. Scatter the objects to be used in this experience over the surface of the sand. Bury them lightly, just enough so they cannot be seen.
2. Divide your class into teams of five or six children and tell them that their teams will be going on a treasure hunt to look for buried treasure. The only rule is that they have to use a magnet to go over the surface of the sand to do their search.
3. As your children find each buried treasure, they should mark it off on their data-capture sheets.
4. When children are finished, all treasure should be rehidden in the sand for the next group.

Extensions
- Ask children whether or not magnets can work in sand. If they are not sure, have them think about how they were able to find their buried treasure. Could they have found their buried treasure without the magnets?
- Can children answer the main question: Will the magnets work through sand?
- If children said "no" at the beginning of this activity, ask them how they were able to find their buried treasure using a magnet if the objects were covered with sand. Would they have been able to find the buried treasure without the magnet?

Closure
Have children complete page 3 of the *Magnetic Fields* do-along book.

Buried Treasure *(cont.)*

Check off all the treasure items you find.

Treasure List	
Picture of Item	Check if Found

45

Water, Water, Everywhere

Question

Will a magnet work through water?

Setting the Stage

Let children know that they will be working with magnets and cups of water.

Materials Needed for Each Individual

- clear plastic cup
- water
- paper clips
- magnets

Procedure

1. Give each child a clear plastic cup half-filled with water.
2. Have each child drop a paper clip in his or her cup.
3. Ask children if there is a way to get the paper clip out of the water without putting their hands in the cup, dumping out the water, or getting their fingers wet.
4. Let children know that they can use a magnet, and they will not even have to get the magnet wet.
5. Give each child a magnet and a little time to play with it.
6. Then, show children by holding the magnet outside the cup and sliding it upward with the paper clip hanging on, they can lift the paper clip out of the water and the magnet is still dry.

Extension

Ask children to answer the main question: Will a magnet work through water? If they cannot, ask them how else this experience could have worked.

Closure

Have children complete page 4 of the *Magnetic Fields* do-along book.

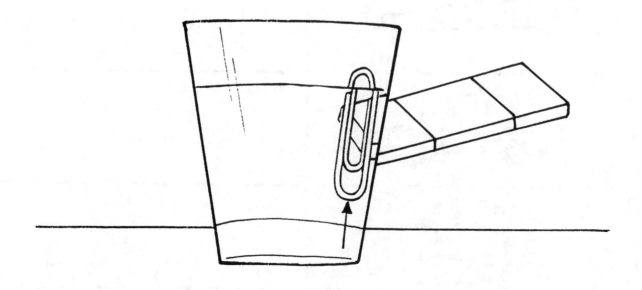

46

Magnetic Force Field

Question

Will a magnet attract a compass needle through a book?

Setting the Stage

Challenge children to make the compass needle move, even though the book is in the way.

Materials Needed for Each Group

- compass
- alnico magnet
- a book

Procedure

1. Have children place the book between the compass and an alnico magnet that is held in one child's hand.
2. Ask children what happens. They should notice that the magnetism goes right through the book as though it were not there and makes the compass needle move.

Extension

Explain to children the magnetic field is strong enough to go through the book. The compass can detect this field and the needle immediately responds to the magnetic field.

Closure

Have children complete page 5 of the *Magnetic Fields* do-along book.

Magnet Trick

Question

Can you make steel balls roll without touching them?

Setting the Stage

Tell children about a trick that they can play on someone by making balls roll without ever touching them.

Materials Needed for Each Group

- cardboard box or plastic dish
- steel balls (ball bearings or BB gun pellets)
- alnico rod magnet

Procedure

1. Place some of the steel balls in each box or dish.
2. Ask one child in each group to carefully hold the box still with his or her hands.
3. Ask another child in each group to pass the alnico rod magnet under the box. Ask what happens. The steel balls should follow the magnet.
4. Ask children to try moving the magnet back and forth or in circles to see what happens.

Extension

Explain to children once again, the magnetic field was strong enough to go through the box and that the steel balls responded to this field.

Closure

Have children complete page 6 of the *Magnetic Fields* do-along book.

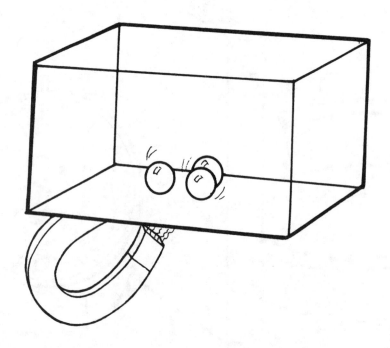

Magnetism Through Gases, Air, and Solids

Question

Will a magnet work through gases, air, and solids?

Setting the Stage

Explain to children the amazing thing about magnets is that they can attract through the strangest things.

Materials Needed for Each Group

- strong refrigerator door magnet
- large bowl
- glass jar
- balloon (uninflated)
- two metal paper clips
- piece of paper
- craft stick
- paper
- about 6" (15 cm) of string
- aluminum foil
- lid from a plastic food container

Procedure

Activity 1 - Magnetism Through Gases

1. Have children tie one end of the string to a paper clip.
2. Have one child hold the other end of the string, as another child slowly moves the magnet toward the paper clip.
3. Then ask your class about the magnetic force that attracts the paper clip. Did the magnet attract the paper clip through the air?

Activity 2 - Magnetism Through Liquids

1. Fill the large bowls half-full with water.
2. Using the paper clip on the string from the first activity, ask children to lower the paper clip into the water.
3. Then tell children to lower the magnet into the water and move it slowly toward the paper clip.
4. Ask children if the magnet attracted the paper clip through the water.

Activity 3 - Magnetism Through a Solid

1. Using the same paper clip on the string, have one child from each group hold the end of the string in the air (just as in the first activity).
2. Ask another child to hold a piece of paper over the front of the magnet and slowly move it towards the paper clip.
3. Ask children if the magnet attracted the paper clip through the paper.

Extension

Have children try these same experiences again, using different solids. They can use a lid from a jar (metal), a craft stick (wood), an uninflated balloon (rubber), a food container lid (plastic), and a drinking glass (glass).

Closure

Have children review their *Magnetic Fields* do-along book.

Just the Facts

Magnets are attracted to some types of materials, but not to others. Most common magnets are made from iron, cobalt, and nickel. Alnico is an alloy of aluminum, nickel, cobalt, and iron. When non-magnetic aluminum is alloyed with the magnetic materials nickel, cobalt, and iron, it makes one of the strongest magnetic materials known.

Aluminum + Nickel + Cobalt + Iron = Alnico

If two north poles are brought close together, they will move away from each other, or repel. If two south poles are brought near each other, they will also repel. If a north pole and a south pole are brought near each other, they will move toward each other, or attract. Opposite or unlike poles attract, but the same or like poles repel.

At very high and very low temperatures, the magnetic qualities of many materials change. Some common non-magnetic materials are wood, aluminum, cloth, and glass. Lines of force will pass easily through these materials. Materials that allow magnetic lines of force to pass through them are called *non-permeable materials.*

Lines of force do not pass through iron or steel, however, which are magnetic materials. The lines of force are gathered in, or absorbed, by the iron or steel. Materials that gather in, or absorb, lines of force instead of allowing them to pass through are called *permeable materials* by scientists.

Permeable materials are often used to protect scientific instruments and watches against the effects of magnetism. The cases of these instruments and watches are made of iron or steel to stop the lines of force from passing through. That way they do not reach the moving parts, attract them, and stop them from moving.

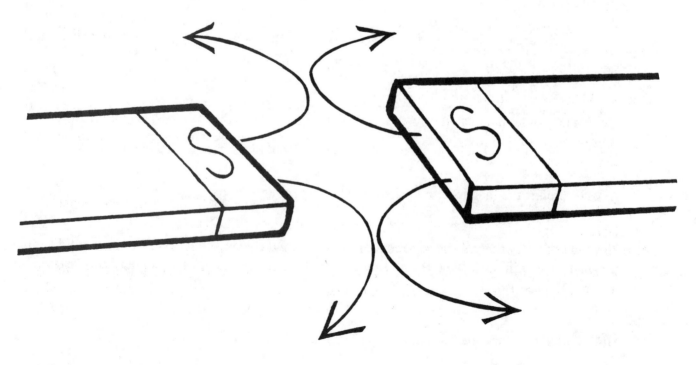

Magnetic Attraction

(For general book-making directions, see page 10.)

Supplies needed for this do-along book:
- crayons
- construction paper
- glue
- silver glitter
- blue glitter
- red glitter
- black glitter
- paper clips
- tacks
- small nails or screws
- scissors
- pennies
- aluminum foil
- scraps of cloth
- wood shavings
- paper

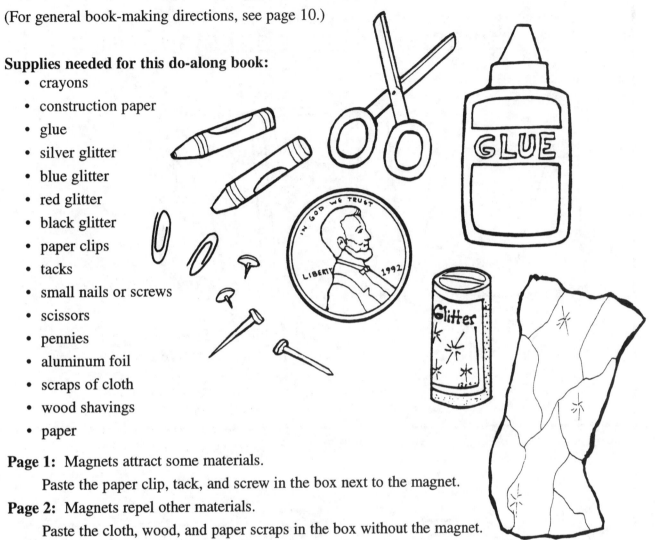

Page 1: Magnets attract some materials.

Paste the paper clip, tack, and screw in the box next to the magnet.

Page 2: Magnets repel other materials.

Paste the cloth, wood, and paper scraps in the box without the magnet.

Page 3: Some metals are not magnetic.

Paste the penny, brass tack, and aluminum foil in the box.

Page 4: Magnets can attract iron filings from sand.

Color the ground tan or gold to represent sand. Color the magnet black. Using a fingertip, lightly spread a thin layer of glue over the area above the sand. Gently shake silver glitter on the glue to represent iron filings.

Page 5: Every magnet has two poles.

Color the magnet black. Using a fingertip, lightly spread glue on the poles of each magnet. Gently shake red glitter on the north poles and blue glitter on the south poles.

Page 6: A magnet's north pole attracts another magnet's south pole.

Color the magnets black. Color the N pole red and the S pole blue. Using a fingertip, lightly spread glue between the magnets. Gently shake black glitter to represent the magnetic field.

Magnetic Attraction (cont.)

2

Magnets repel other materials.

Magnets attract some materials.

1

Magnetic Attraction (cont.)

Magnets can attract iron filings from sand.

Some metals are not magnetic.

3

Magnetic Attraction (cont.)

9

A magnet's north pole attracts another
magnet's south pole.

Every magnet has two poles.

5

Magnetic Walk

Question

What things outside the classroom will be attracted to magnets?

Setting the Stage

Let children know that they are going for a walk around the schoolyard to find things that are attracted to magnets.

Materials Needed for Each Group

- magnets (1 per 2-3 children in a group)
- large chart paper or butcher paper
- colored markers or crayons

Procedure

1. Take children for a walk around the school. Tell them to look for things that are attracted to magnets. If they can pick up the items they find, they can bring them back to class and share them with the whole class.

2. When the walk is over, help children draw a large magnet on the chart or butcher paper and then list all the things that were attracted to their magnets.

3. After each group has finished its chart, have children who brought items back to class share them with everyone.

Extension

Children can look for things at home that would be attracted to a magnet. Remind them to keep magnets away from computers, videos, and software.

Closure

Have children complete pages 1 and 2 of the *Magnetic Attraction* do-along book.

Sand Trap

Question

Can a magnet find iron filings in sand?

Setting the Stage

Tell children that they will see if they can find magnetic material in sand.

Materials Needed for Each Group

- iron filings
- bar magnet
- horseshoe magnet
- baby food jar
- sheet of construction paper

Procedure

1. Take children to the playground sandbox with a large magnet and have them take turns running the magnet through the sand.
2. Natural iron filings will collect on the magnet; help your children put them into the baby food jar.
3. Help children sprinkle some iron filings onto a sheet of construction paper.
4. Tell children to place the bar magnet on the filings and move it around several times.
5. Then tell children to do the same with the horseshoe magnet. (The filings will collect at the ends.)

Extension

Explain to children the poles of the magnets are the strongest parts and therefore will collect the most filings.

Closure

Have children complete pages 3 and 4 of the *Magnetic Attraction* do-along book.

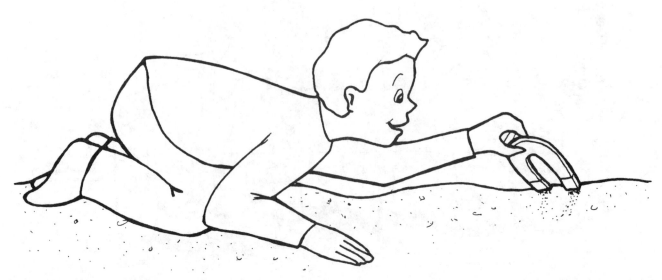

North Pole, South Pole

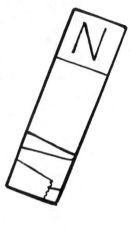

Question
Can you find the north and south poles of a magnet?

Setting the Stage
Tell children that they will use magnets to determine their poles.

Materials Needed for Each Group

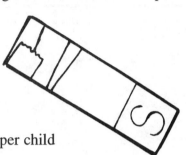

- two bar magnets
- red and blue tape
- blue crayon
- red crayon
- data-capture sheet (page 58), one per child

Procedure

1. Place two bar magnets on each table. Put red tape around the south poles and blue tape around the north poles of each magnet.

2. Explain to children sometimes a magnet can pick up another magnet and sometimes a magnet can push another magnet away. Explain your color coding of the poles.

3. Tell children to look at the two bar magnets on their table. Tell them to put the two north poles together. Then have them draw a picture on their data-capture sheets of what happened, coloring the north poles blue.

4. Then have children put a north pole and south pole together. Then have them draw a picture on their data-capture sheets of what happened, coloring the poles red or blue depending on their side.

5. Finally have children put two south poles together. Then have them draw a picture on their data-capture sheets of what happened, coloring the south poles red.

Extensions
- Tell children to cross the bar magnets, one atop the other. Ask them what happens.
- Have children complete the activities on pages 59-60.

Closure
Have children complete pages 5 and 6 of the *Magnetic Attraction* do-along book.

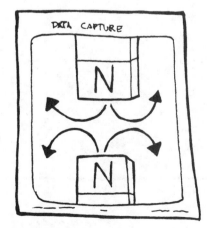

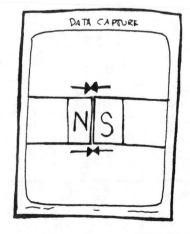

North Pole, South Pole *(cont.)*

Fill in the information.

1. Put the two north poles together. Draw a picture of what happens, coloring the north poles blue in your picture.

2. Put a north pole and south pole together. Draw a picture of what happens, coloring the poles red or blue in your picture.

3. Put two south poles together. Draw a picture of what happens, coloring the south poles red in your picture.

North Pole, South Pole (cont.)

Find five objects that might be picked up by a magnet.   Write the name of each object in the "YES" column.

Find five objects that might not be picked up by a magnet.   Write the name of each object in the "NO" column.

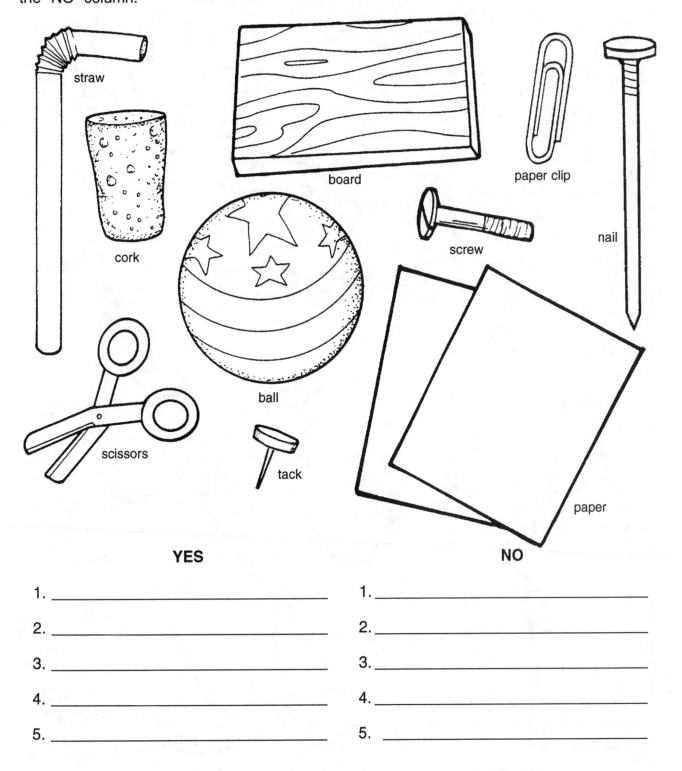

straw

cork

board

paper clip

nail

screw

ball

scissors

tack

paper

YES	NO
1. _____	1. _____
2. _____	2. _____
3. _____	3. _____
4. _____	4. _____
5. _____	5. _____

North Pole, South Pole *(cont.)*

Put an x on each object a magnet will pick up.

60

Just the Facts

Have you ever wondered what kinds of things you can do with magnets? Magnets can help us at home in many different ways. With a decorative covering, magnets can hold schoolwork on the refrigerator. They can also be useful when picking up small items, such as sewing pins. At school, magnets can hold to white boards, connect plastic blocks, be placed on the back of small letters to help the young learners to spell, and be used for many other tasks.

In the industrial world, huge magnets help the automobile salvage companies to move large vehicles. Magnets are also used in all electric motors and in generators that produce electricity. They help produce the pictures on television and computer screens. Magnets are used in doorbells, car ignitions, tape players, video cassette recorders, record players, microphones, telephones, speakers, electric guitars, and in many other pieces of equipment.

Magnets are useful on farms, too. Farmers use them in their tractors and other machinery.

Magnets have become very useful in our everyday lives.

We Use Magnets Every Day

(For general book-making directions, see page 10.)

Supplies needed for this do-along book:
- crayons
- construction paper
- glue
- scissors
- pennies
- scraps of cloth
- small squares of cardboard 1/2" x 1/2" (1.25 cm x 1.25 cm)
- gold button
- silver buttons
- paper clips

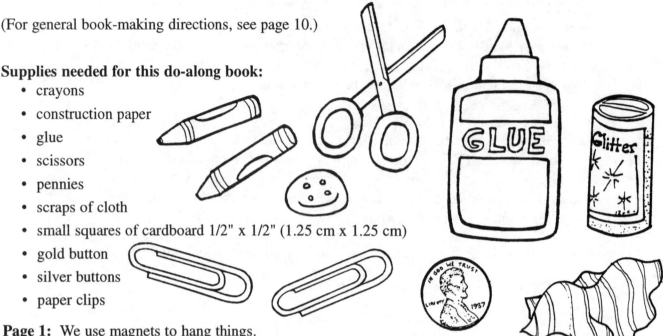

Page 1: We use magnets to hang things.

Color the refrigerator and stove any color. Then glue a scrap of paper on the refrigerator and a small square of cardboard on top of the paper. Glue the scrap of cloth on the stove and a little colored square on top of that.

Page 2: We use magnets in transportation.

Color the road area black. Color the car, train, bus, and airplane.

Page 3: We use magnets in the kitchen.

Color the can opener and paste a silver button on the can opener to represent the magnet.

Page 4: We use magnets in the living room.

Color the door any color, and color the stereo brown. Paste a gold button on the door and paste a silver button on the stereo motor.

Page 5: We use magnets in industry.

Color the buildings different colors. Color the smoke gray. Glue two silver buttons on top of the steel mill to represent magnets inside the mill.

Page 6: We use magnets in communication.

Color the telephone and paste a silver button on each end of the receiver to represent the magnets.

We Use Magnets Every Day (cont.)

We use magnets in transportation.

We use magnets to hang things.

We Use Magnets Every Day (cont.)

We use magnets in the living room.

4

We use magnets in the kitchen.

3

64

We Use Magnets Every Day (cont.)

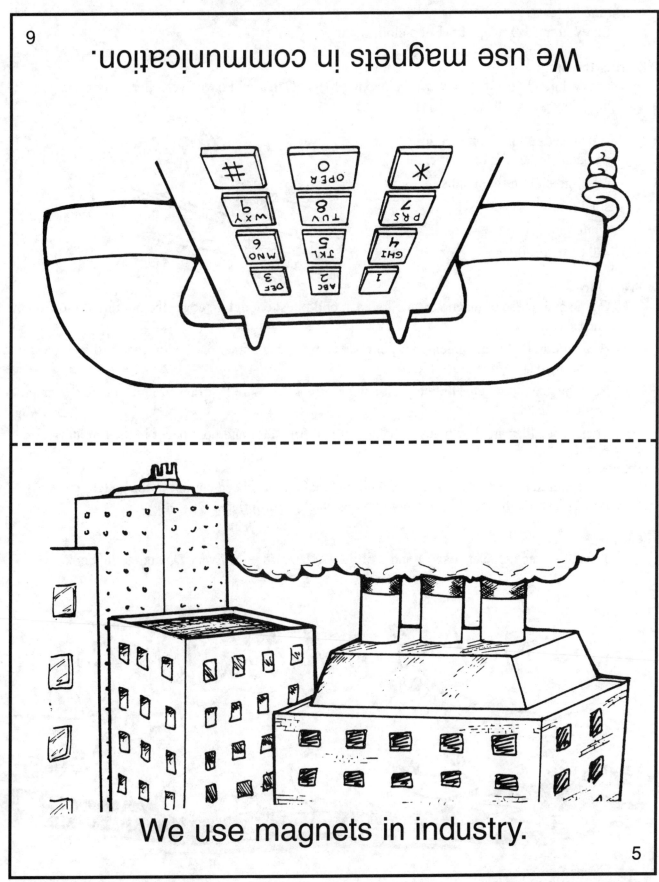

We use magnets in communication.

We use magnets in industry.

Crocodile Magnet

Question

Can you make a magnet out of a clothespin?

Setting the Stage

Let children know that they will be making magnets that will have eyes. See if they can guess how they will do this.

Materials Needed for Each Group

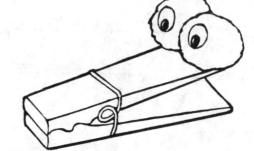

- wooden clothespin, one per child
- small green pom-pom balls
- small plastic wiggle eyes
- white glue
- roll of magnetic tape

Procedure

1. Cut a small piece of magnetic tape for each child, just enough to cover the bottom of the clothespin.
2. Have children glue the green pom-pom balls next to each other on the top side of the clothespin to create the back of the crocodile.
3. Next, have children glue on the plastic wiggle eyes. Be sure that they are glued onto the green pompom balls.
4. Their crocodiles are now ready to hang on the refrigerator and hold paper in their mouths.

Extension

Have children create other magnets with material that has been cut out in the pattern of butterflies, or any pattern that would be seasonally appropriate.

Closure

Have children complete page 1 and 2 of the *We Use Magnets Every Day* do-along book.

Magnet Mobile

Question

Can you make a mobile without using any glue?

Setting the Stage

Tell children that they will be reviewing the different types of magnets by cutting out patterns and making a mobile with them.

Materials Needed for Each Group

- copy of patterns (page 68), one per child
- roll of magnetic tape (has a sticky side and a non-sticky side)
- string
- clothes hanger (one per child)
- box of paper clips

Procedure

1. Have children cut out the patterns.
2. Next, cut a small piece of the magnetic tape for each child, pull off the backing, and place the sticky side on the pattern.
3. Have children cut pieces of string and tie one end to the clothes hanger and attach a paper clip to the other end.
4. Have children place the paper clip on the magnetic side of the pattern. The pattern will then hang from the paper clip. Do this to all the patterns.

Extension

Patterns can be made for any unit being studied or for seasonal activities.

Closure

Have children complete page 2 of the *We Use Magnets Every Day* do-along book.

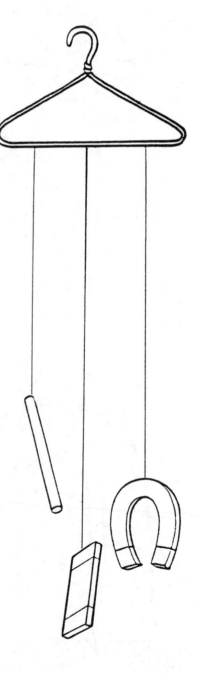

Magnet Mobile

Cut out patterns.

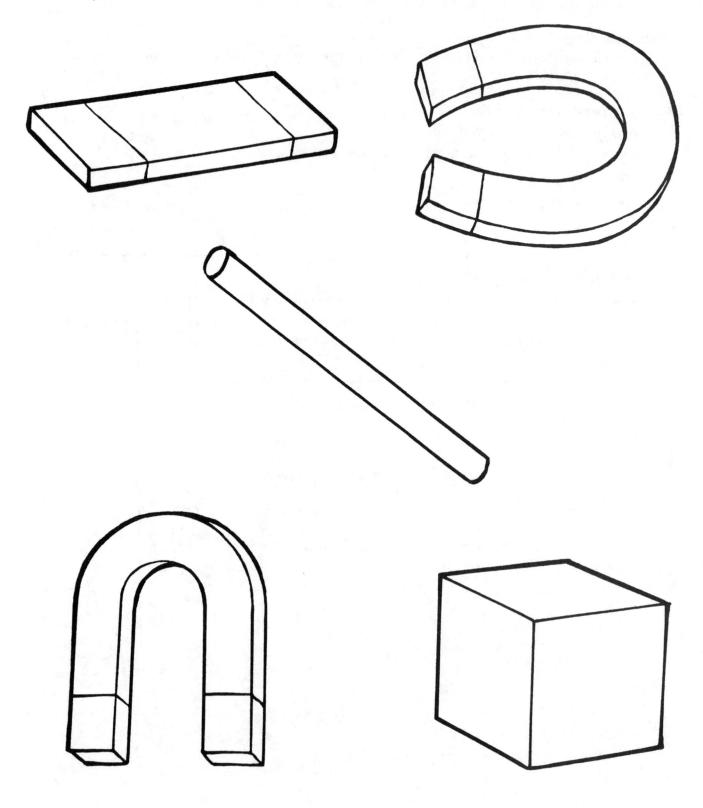

68

Magnets in the Home

Question

Can you find six ways you and your family use magnets at your house?

Setting the Stage

Tell children that this activity will get their families involved with a science activity at home.

Materials Needed for Each Individual

- pencil or crayon
- data-capture sheet (page 70), one per child

Procedure

1. Assign children the task of looking for magnet uses in the home.
2. Provide a data-capture sheet (page 70) for recording their information. Tell children to have their parents help them fill it out.

Extension

After children bring their results back, they can share their responses with the class. Hang their data-capture sheets on the bulletin board or around the classroom. After all the data is collected, find the most common occurrences (on the refrigerator, on the stove, etc.) and help children to create a class graph showing how many children found magnets in those areas.

Closure

Have children complete pages 3 and 4 of the *We Use Magnets Every Day* do-along book.

Magnets in the Home *(cont.)*

How many magnets can you find in your house?

Parent Instructions: Help your child fill in the blanks with the number of magnets found.

In the squares below, draw a picture of each magnet you find.

I found _____ magnets in the kitchen.

I found _____ magnets in the bathroom.

I found _____ magnets outside.

I found _____ magnets in my bedroom.

I found a total of _____ magnets.

Complete pages 5 and 6 of the *We Use Magnets Every Day* do-along book.

Language Arts

Science Concept: Magnets are used around the world.

Have an oral discussion about magnets. Write down a sentence from each child. Compile these into a story and read it back to them. Give each child a copy of the story.

Math

Science Concept: Magnets have many uses.

The Venn diagram is a way of visualizing sets of objects.

To use a Venn diagram, you must have at least three things—two sets of things with distinguishable characteristics, and a third group (drawn from the first two) which contains the same characteristic.

For a class that is unfamiliar with Venn diagrams, you can start with something with which they can identify. Draw two large circles on the board or on large chart or butcher paper. Put these labels in the circles:

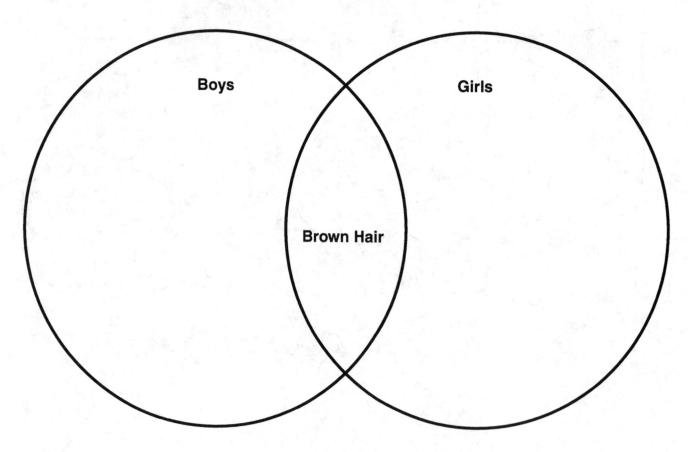

Each child can identify his or her position within the circle by placing his or her name in the category that pertains to him or her.

If a boy or girl does not have brown hair, he or she would place his or her name in one of the outside circles.

Math *(cont.)*

Here is a Venn diagram for magnets.

Materials Needed
- two lengths of light rope or cord, 15' (5 m) each
- three pre-lettered signs: "Magnetic," "Non-Magnetic," and "Metal"
- several magnets
- collection of objects: cotton balls, copper wire, paper towels, rubber erasers, plastic cups, iron nails, aluminum foil, pencils, paper fasteners, crayons, paper clips, etc.

Procedure
1. Overlap two large circles of light rope or cord on the classroom floor.
2. In separate circles, place the signs saying "Magnetic" and "Non-Magnetic."
3. In the overlapping area, place the sign saying "Metal."
4. With the collection of objects on a nearby desk, model this activity by using a magnet to test a blackboard eraser, a paper hole-punch, and an aluminum pie-plate, before placing each in its proper area.
5. Have children test all collected objects and place them in the correct area of the Venn diagram.

   Note: This activity may be repeated from time to time with different objects until the children become confident and adept with the process of classifying. It is a hands-on tactile experience that promotes learning and retention.

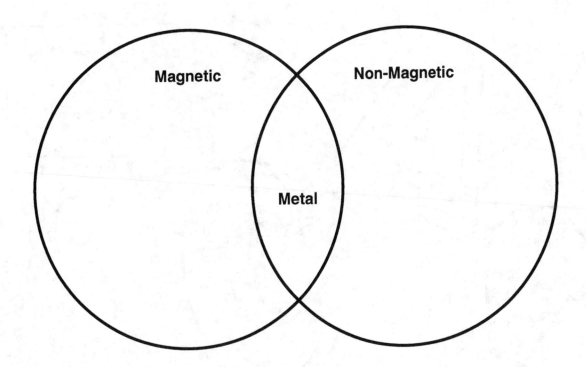

Social Studies

Science Concept: Uses of magnets in the home.

Have children find or bring in samples of magnets they have at home or pictures found in magazines.

They can share these with the class during show and tell.

Social Studies *(cont.)*

Science Concept: How people use magnets.

Have children investigate the ways people use different magnets. The children can create a mural of all their discoveries.

Art

Science Concept: Magnets can be used doing everyday tasks.

Have children try using magnets to paint.

1. Put a few objects made of iron or steel in some tempera paint.
2. Have children cut a piece of white paper to fit the bottom of a shallow glass baking dish.
3. Tell children to put the paper in the bottom of the glass dish.
4. Then have children put the paint-coated objects into the dish.
5. Next, have each child take turns being a magnet artist and move a magnet under the dish to pull the paint-coated iron and steel objects around on the paper to make a painting.

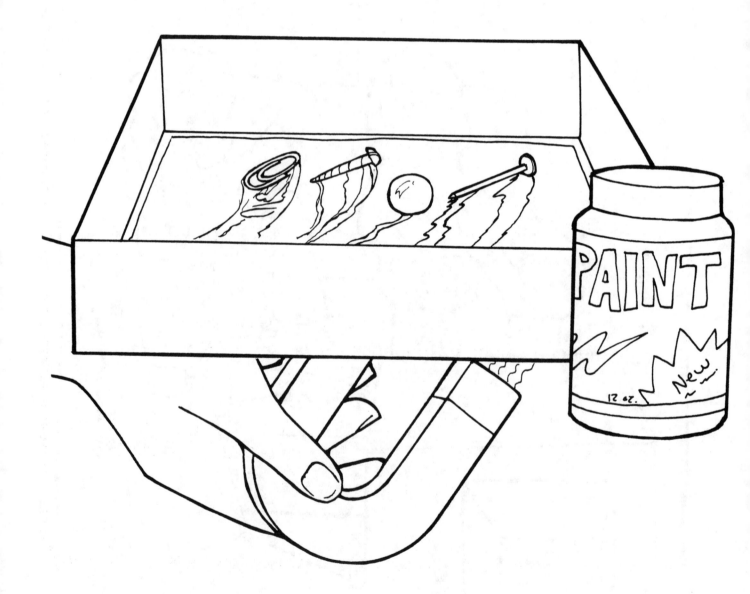

76

Science Safety

Discuss the necessity for science safety rules. Reinforce the rules on this page or adapt them to meet the needs of your classroom. You may wish to read them to the children often for reinforcement.

1. Begin science activities only after all directions have been given.

2. Never put anything in your mouth unless it is required by the science experience.

3. Always wear safety goggles when participating in any lab experience.

4. Dispose of waste and recyclables in proper containers.

5. Follow classroom rules of behavior while participating In science experiences.

6. Review your basic class safety rules every time you conduct a science experience.

You can still have fun and be safe at the same time!

Data-Capture Sheet

One of the greatest frustrations for teachers with young learners involved in hands-on experiences is trying to gather (capture) what children have learned. It is often difficult and time-consuming to meet with all children to have them share their experiences. Have children periodically complete a data-capture sheet. Design sheets so that children need to mainly draw and write inventive or simple sentences about what they observed.

Blank Data-Capture Sheet

Assessment Strategies

Science Artwork

Young children love to draw. "Pictures speak a thousand words." It is appropriate, and acceptable, to utilize children's artwork as a means of assessment. An excellent year-long science artwork assessment tool is a Science Art Journal which will contain artwork from all topics learned during science time.

To make an artwork journal, begin by punching three holes on the short left side of two 12" x 18" (30 cm x 45 cm) pieces of tagboard and ten to fifteen sheets of cut 12" x 18" (30 cm x 45 cm) chart/butcher paper. Create a book by sandwiching the chart/butcher sheets between the tagboard cover sheets. Line up the left short side and punch three holes (top, middle, bottom) through entire thickness. Thread a short piece of string through each hole and tie in a knot or bow (make certain that the pages can turn easily before knotting tightly), or alternatively, open three small metal clasp rings and place through the three holes and shut rings. Write "The 'Art' of Science" and the child's name on the cover. When the child has completed the first piece of artwork you want to place in the journal, simply glue or tape the artwork to the first sheet of chart/butcher paper. Then place the date in the top corner and write any comments you and/or the child desires to make pertaining to the piece of artwork.

This is an excellent tool to use during parent-teacher conferences. It is also a great resource for the child's next teacher. This type of visual log can follow the growth that took place during a number of years if the continuing teacher simply adds more pages to the back of the book.

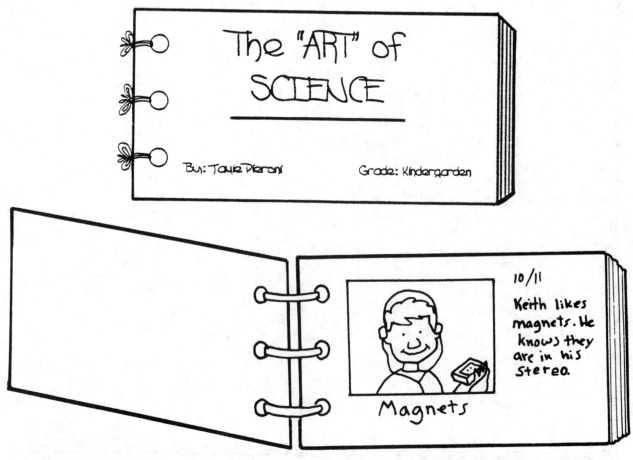

Assessment Strategies *(cont.)*

Science Projects

Provide opportunities throughout the year for children to create projects based on their science topics or themes. Among the most popular for the young learner are these:

Posters

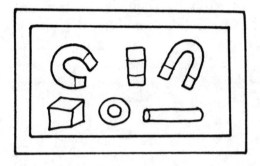

Dioramas

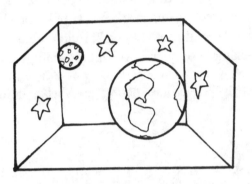

Mobiles

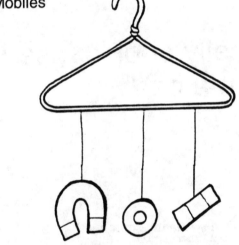

Student-Created
Bulletin Boards

Scientist at Work (Photographs)

Keep a camera in the classroom and regularly take pictures of "learning in progress." Keep the pictures in a photographic time line to refer to during assessment and evaluation processing.

Assessment Strategies *(cont.)*

The "Why" Game

Nonthreatening assessments are in popular demand. This term means that children do not have to "worry" about a right/wrong factor in their responses. The "Why" game can be oral and/or written. The basic concept of this assessment is that, given a specific statement, there are two extreme responses with a continuum between them. Neither an extreme response nor any response in between is correct or incorrect; it is simply the child's choice. The only responsibility a child has, once a choice has been made, is to back it up with a reason (facts + personal experience = reason). Below is an example of what two children said given the statement, "A magnet is . . ."

Billy (Lived in Anchorage, Alaska, before moving to your school site.)

A   is . . .   1   2   3   4   6   7   8   9   10.

                                             small   big

(Scripted by the teacher):

"I think that magnets are really cool 'cause I watched one lift up a car."

Sally (Spent the last two summers with her Aunt on the Cape Cod's Atlantic shoreline.)

A   is . . .   1   2   3   4   6   7   8   9   10.

                                               small   big

(Scripted by the teacher):

"Magnets are fun. They helped me find a ring at the beach."

Assessment Strategies *(cont.)*

Science Portfolios

A portfolio is a collection of a given child's work and teacher-collected data. The collecting of the work and data begins at the start of the school year. The portfolio is to be evaluated periodically as the year unfolds. (If you have grading periods, that is the optimum time for evaluation.) To avoid being overwhelmed by a thickening portfolio as the year goes by, place a piece of colored paper after the evaluated materials to signify the end of that assessment period and the beginning of another. This way you will not have to go back to the beginning each time the next evaluation process begins.

A teacher may include samples generated from artwork, projects, photographs, story writing, and "Why" game responses as discussed on the previous assessment pages, along with a variety of teacher-collected data. Teacher-collected data may be data forms provided by the school system or school site. Two alternative teacher-collected data resources are outlined below.

Science Progress Report

On the next page you will find a blank progress report. This report can be used for each grading period as a method of keeping parents, as well as the next year's teacher (if reports are passed on at the close of the school year), informed.

On the progress report, the academic development section has boxes for up to eight grading periods. Each box is divided into three sections. The top left section of the box is used to write in the instructional level (Y, at year: Y+, above year; Y-, below year). The top right box is used to write in the student progress (M+, 90% - 100% mastery; M, 80% - 89% mastery; I, incomplete or working towards mastery). The bottom half of the box is used to record the behavior science statements. Ten statements are listed at the bottom of the report. You are encouraged to use more than one statement per box.

Conferencing Form

The conferencing form (blank, page 85) works especially well with the nonwriters (those who have not even begun inventive spelling). The teacher simply acts as the recorder and logs in both the questions and responses. The teacher can then add the comments immediately after the conferencing time or later in the day or week.

Science Progress Report

Student's Name _____ Year/Grade _____

Instructional Level

Y   at year

Y+   above year

Y-   below year

Process

M+   90-100% mastery on assessments

M   80-89% mastery on assessments

I   incomplete — working toward mastery

Example

Date:

Date:

Date:

Date:

Date:

Date:

Date:

Date:

Behavior Statements

1. Demonstrates outstanding effort
2. Demonstrates satisfactory effort
3. Demonstrates inconsistent effort
4. Completes work on time
5. Needs to improve quality of work

6. Assignments often incomplete or late
7. Participates in more challenging activities and extensions
8. Making progress toward mastery
9. Often needs reteaching and retesting
10. Progress is hindered by frequent absences

Conferencing Form

Student Conference Record for Science

Date: _____

Scientist's Name: _____

Teacher Questions	Student Response

Teacher Comments: _____

_____

_____

_____

Science Observation Journal

Cover Sheet

Name _____

86

Science Observation Journal

Blank Page

Here is a picture of what I saw.

Letter to Parents

What Does Science Do for My Child?

Science stimulates a child to

. . . explore and experiment.

. . . use a variety of materials for learning.

. . . develop skills in using the scientific method for problem solving, observing, identifying, predicting, and generalizing.

. . . work cooperatively with others.

. . . verbally share with others what has been discovered.

. . . write down or draw what has been learned.

. . . move from concrete to abstract thinking by being involved in hands-on experiences.

. . . discover the life, earth, and physical world around him/her.

Parental Assistance Form

Parents can be a big help on a class field trip or as assistants in small group experiences in the classroom. They reduce the student-adult ratio significantly and increase the personal attention available to students.

Use the forms below to elicit the assistance you may need in the classroom or on field trips.

Dear Parents,

We will be needing some parents to volunteer for our field trip on

_____ to _____.

We will be leaving school at _____ and will return

about _____.

If you would like to be a chaperone, please send a return note to school by

_____.

Sincerely,

_____

Teacher

SCHOOL BUS

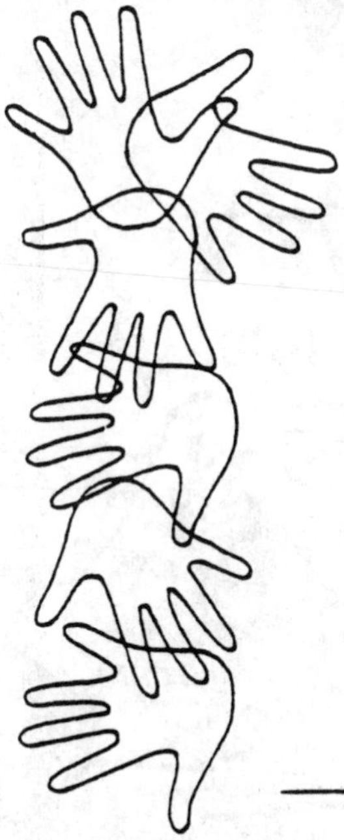

WANTED!
Helping Hands

Date:

Time:

To Help With:

Please return this note with your signature if you can help with this activity.

I can help on _____ at _____

_____

Signature and Phone Number

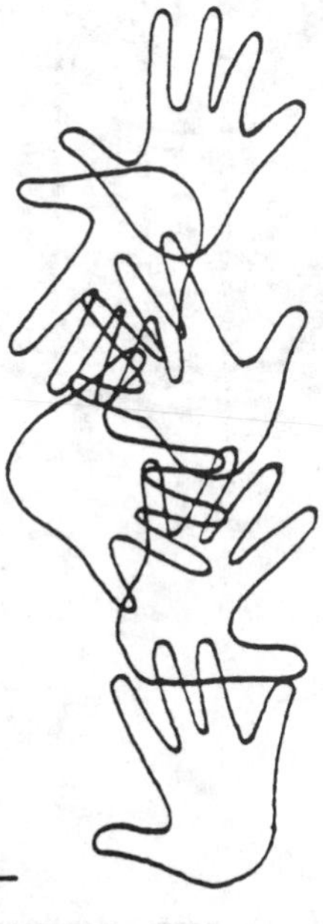

Messy Day Letter

Use these reminder slips when discovery experience days will be very hands-on.

We Need To Dress for Mess

on _____
    Date

when we will be _____
                Messy Activity

_____.

A Messy Day is Coming Our Way

on _____
    Date

when we will be

_____.
            Messy Activity

90

Supply Request Letter

Dear Parents,

Our class will be involved in many science experiences this month. Could you help to make these experiences a success by looking around the house for the items listed below? If so, please send them in with your child as soon as possible.

Thank you!

We're Going on a Field Trip

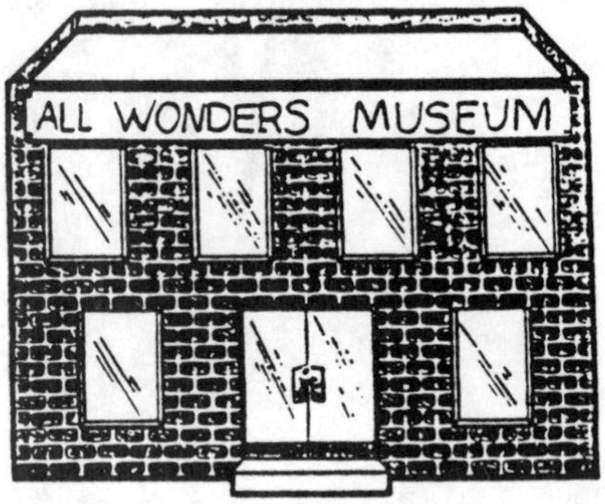

ALL WONDERS MUSEUM

Where: _____

_____

When: _____

_____

Why: _____

_____

How: _____

Please let me bring: _____

_____

Please sign the permission slip **below** and have your child return it
by _____. Your child will not be allowed to participate
without the signed slip.

Thank you.

_____
Teacher

- -

My child, _____ ,

has my permission to participate on **the science experience** to

_____ .

In case of emergency contact _____ .

Phone: _____

MUSEUM

_____
Parent

☐ I will be able to chaperone. **Please contact me!** _____

Daytime Phone Number

Science Award

An Awesome Award

given to

for being a
magnet lover!

Teacher

Date

Glossary

Attraction—when two charges or poles are different, they "attract" or come together.

Atom—the smallest part of a substance. Inside the atom you will find the electrons, neutrons, and protons.

Compass—an instrument for determining directions by the pointing to the earth's magnetic north.

Conclusion—the outcome of an investigation.

Electron—a particle inside an atom that carries a negative charge.

Experiment—a means of proving or disproving an hypothesis.

Geographic North—the earth's geographic North Pole.

Hypothesis—an educated guess to a question which one is trying to answer.

Interaction—reciprocal action or influence.

Investigation—observation of something followed by a systematic inquiry in order to understand what was originally observed.

Lodestone—a variety of magnetite that shows polarity and acts like a magnet when freely suspended.

Magnet—an object that has a magnetic field around it.

Magnetic Field—the area around a magnet that causes magnetic movement.

Magnetic Needle—a freely movable, needle-shaped piece of magnetized material which tends to point to the north and south (magnetic) poles of the earth.

Magnetic North—the earth's magnetic north pole. This pole continually changes with the earth's magnetic field.

Magnetic Pole—the ends of a magnet. One pole is north, and one pole is south.

Magnetism—an invisible force that can make objects move away, move together, or stay in the same place.

Molecule—the smallest part of an element, substance, or compound that can exist freely in the solid, liquid, or gaseous state and still retain its composition and properties.

Navigator—One who directs the course of a ship, aircraft, etc.

Neutron—a particle inside an atom that carries a neutral charge.

Glossary *(cont.)*

Observation—careful notice or examination of something.

Permeable—allowing passage, especially by fluids, but also pertaining to absorption of magnetic fields.

Procedure—a series of steps that is carried out when doing an experiment.

Proton—a particle inside an atom that carries a positive electrical charge.

Question—a formal way of inquiring about a particular topic.

Repel—to push away. When two charges or poles are the same, they repel each other.

Resistance—opposition to the flow of electrons.

Results—the data collected after performing an experiment.

Scientific Method—a creative and systematic process of proving or disproving a given question, following an observation. Observation, question, hypothesis, procedure, results, conclusion, and future investigations comprise the scientific method.

Scientific-Process Skills—the skills necessary to have in order to be able to think critically. Process skills include: observing, communicating, comparing, ordering, categorizing, relating, inferring, and applying.

Scientist—a person considered an expert in one or more areas of science.

Variable—the changing factor of an experiment.

Bibliography

Adler, David. *Amazing Magnets.* Troll Associates, 1983.

Amery, H. & A. Littler. *The Know How Book of Batteries and Magnets: Safe and Simple Experiments.* Educational Development, 1977.

Ardley, Neil. *Science Book of Magnets.* HBJ, 1991.

Berman, Paul & Keith Wicks. *Science in Action* (6 volumes). Marshall Cavendish Corp, 1988.

Brown, Robert. *333 Science Tricks and Experiments.* Tab Books, Inc., 1984.

Burton, Virginia Lee. *Mike Mulligan & His Steam Shovel.* Houghton & Mifflin, 1939.

Catherall, Ed. *Exploring Magnets.* Steck-Vaughn Library, 1990.

Challand, Helen, J. *Experiments with Magnets.* Childrens Press, 1986.

Challand, Helen, J. *The New True Book of Experiments with Magnets.* Childrens Press, 1986.

Cooper, Jason. *Magnetism.* Rourke Corp, 1993.

Davis, Kay & Wendy Oldsfield. *Electricity & Magnetism.* Raintree Steck-V, 1991.

Guthridge, Sue. *Thomas A. Edison: Young Inventor.* Macmillan, 1986.

Henry, Lucia K. *Science & Ourselves.* Fearson Teach Aids, 1989.

Hoyt, Marie A. *Magnet Magic Etc.* Educ Serv Pr., 1983.

Jennings, Terry. *Electricity & Magnetism.* Childrens, 1989.

Jennings, Terry. *Magnets.* Watts, 1990.

Kaufman, Mervyn. *Thomas Alva Edison: Miracle Worker.* Chelsea Hsa, 1993.

Kirkpatrick, Rena. *Look at Magnets.* Raintree, 1978.

Macaulay, David. *The Way Things Work.* HM, 1988.

Parker, Steve. *Thomas Edison & Electricity.* Harp C Child Books, 1992.

Santrey, Laurence. *Magnets.* Troll Assocs., 1985.

Simon, Seymour. *Einstein Anderson, Science Sleuth.* Puffin Bks., 1986.

Vivian, Charles. *Science Experiments & Amusements for Children.* Dover, 1967.

Ward, Alan. *Forces and Energy.* Watts, 1992.

Ward, Alan. *Magnets and Electricity.* Watts, 1992.

Weinberg, Michael. *Thomas Edison.* Longmeadow Press, 1988.

Whyman, Kathryn. *Electricity & Magnetism.* Gloucester, 1986.

Wood, Robert W. *Science for Kids.* Tab Books, 1992.

Technology

Agency for Instructional Technology. *Magnetism: Why Does a Compass Point North?, Electricity: Where Does Electricity Come From?,* and *Let's Explore Magnets.* Available from ATI The Learning Source, (800) 457-4509. video

Bill Walker Productions. *Electricity and Magnets.* Available from Cornet/MTI Film & Video, (800) 777-8100. film, video, and videodisc